Gene Kizer writes with authority from the desire to tell the truth. His common sense style is the product of honesty. One cannot read his work without concluding that this is a man to be trusted.

James Everett Kibler

Critically acclaimed novelist,
poet and scholar

February 17, 2015

Dr. Kibler is author of *Our Fathers' Fields*; *Memory's Keep*; *Walking Toward Home; The Education of Chauncey Doolittle; Tiller;* and *The Gentles Gamester*. He was an English professor at the University of Georgia for many years.

Other Books by Gene Kizer, Jr.

Slavery Was Not the Cause of the War Between the States

The Irrefutable Argument.

The Elements of Academic Success

How to Graduate Magna Cum Laude from College (or how to just graduate, PERIOD!)

Charleston, SC Short Stories, Book One

Six Tales of Courage, Love, the War Between the States, Satire, Ghosts & Horror from the Holy City

Charles W. Ramsdell

Dean of Southern Historians

Charleston Athenaeum Press

Charles W. Ramsdell

Dean of Southern Historians

Volume One:

His Best Work

Compiled,

and with Introduction by

Gene Kizer, Jr.

CHARLESTON ATHENAEUM PRESS

Charleston and James Island, S.C.

Charleston Athenaeum Press
Bonnie Blue Publishing

www.CharlestonAthenaeumPress.com
www.BonnieBluePublishing.com

Post Office Box 13012
Charleston, South Carolina 29422-3012

ORDER SIGNED COPIES FROM
www.BonnieBluePublishing.com

ISBN: 978-0-9853632-3-9 *(softcover)*

First Print Edition
May 10, 2017

Charles William Ramsdell, Ph.D.

With great pleasure,
to the Usual Suspects
and some new ones

Dad; Mom, *Rest in Peace*;
My Children, Trey and Travis,
and to Trey's Heather, and to
Heather's family: Larry Woodward
and Holly Morgan Woodward,
and Alana Woodward, and
Carrie Jones Stephens;
Aunt Betty Ann Kizer, and
Cousin Paula M. Weatherford;
Clay and Kim Martin, and Clay's dad,
CDR James Richard Martin, USN,
***Rest in Peace*, and Kim's mom,**
Hazel Thomas, *Rest in Peace*;
Capt. Richard Sharpe;
Sgt. Patrick Harper

And to the Land of the Free,
And Home of the Brave.

Contents

More Treatises

Ramsdell's In-Depth Analyses of Challenges Behind Confederate Lines that Greatly Affected the War

Ramsdell's Book Reviews
Are Works of Art

Bibliography of Ramsdell's Writings

Introduction

"In all that pertained to the history
of the Southern Confederacy, his scholarship
was decisive."[1]

In Memoriam
Charles William Ramsdell
University of Texas

I am deeply honored to bring out the writings of one of the greatest Southern historians of the first half of the twentieth century, Charles W. Ramsdell (1877-1942). His well-deserved title, Dean of Southern Historians, was given to him by his peers to acknowledge his scholarship and stature as the primary authority of his time on the Confederate States of America and much of Southern history.

He was a Texan and quintessential Southerner and saw things through those eyes. Objectivity, evidence and rigorous argument were the sacred standard for historians back then. It wasn't always attained but it was a far better standard than the political correctness of today. Ramsdell was analytical and known for sound judgment, and he wrote with clear vivid prose that is easy to read and comprehend.

---

[1] *In Memoriam, Charles William Ramsdell*, Index of Memorial Resolutions and Biographical Sketches, The University of Texas at Austin, https://wikis.utexas.edu/display/facultycouncil/Memorial+Resolutions, accessed November 29, 2016.

Professor Ramsdell taught at the University of Texas at Austin most of his long career. He held "visitor lectureships in the state universities of Illinois, Colorado, West Virginia, Missouri, North Carolina and Louisiana; and in Columbia, Northwestern, Western Reserve and Duke Universities."[2] Ramsdell's papers are at UT's Dolph Briscoe Center for American History and include in a Biographical Note: "Recognized as the dean of Southern historians, Dr. Ramsdell held the distinction of being the most distinguished scholar and teacher in the field of Southern history."[3] There is still today "The Fletcher M. Green and Charles W. Ramsdell Award" given by the Southern Historical Association for the "best article published in the *Journal of Southern History* during the two-preceding years."[4]

I have left the details of Ramsdell's life out of this Introduction because they are included in the first treatise in this book, "Charles W. Ramsdell: Historian of the Confederacy," by Wendell Holmes Stephenson, a distinguished historian himself and colleague of Ramsdell.

It is highly beneficial in this day and age to study the

---

[2] Ramsdell, Charles W., short biography on Texas State Historical Association website by J. Horace Bass, https://tshaonline.org/handbook/online/articles/fra25, accessed October 25, 2016.

[3] Biographical Note in *A Guide to the Charles Ramsdell Papers, 1844-1942*, Dolph Briscoe Center for American History. http://www.lib.utexas.edu/taro/utcah/01314/cah-01314.html, accessed October 20, 2016.

[4] Southern Historical Association website, http://thesha.org/awards/ramsdell, accessed October 25, 2016.

writings of renowned historians prior to 1960, especially Southern historians. They knew almost as much as historians today — certainly they knew all the major issues and arguments of American history — but they were not corrupted by political correctness. They were interested in a broad narrative of our great country and its part in Western Civilization. Since 1960, the racist identity politics of the left has degraded American history, especially in academia.

One of the problems with academia is that, in a metaphorical sense, it is inbred.

It is so liberal, the 33 wealthiest colleges in the United States gave Hillary Clinton $1,560,000. They gave Donald Trump $3,000.[5]

Over 90% of professors in the humanities and social sciences, which include history, are liberals, and it has been this way for decades.[6] Those with differing opinions,

---

[5] The 33 wealthiest colleges in the United States also gave Bernie Sanders $648,382, so, adding Hillary Clinton's $1,560,000 to Bernie's $648,382 gives a wopping $2,208,382 that academia gave to two extremely liberal Democrat candidates (99.9%) while giving $3,000 to Donald J. Trump (.136%), who won the presidency. See "Donald Trump Campaign Lacking In Support From Academic Donors" by Carter Coudriet, August 16, 2016, http://www.forbes.com/sites/cartercoudriet/2016/06/16/donald-trump-campaign-lacking-in-support-from-academic-donors, accessed January 25, 2017.

[6] See Horowitz, David and Jacob Laksin, *One-Party Classroom: How Radical Professors at America's Top Colleges Indoctrinate Students and Undermine Our Democracy* (New York: Crown Forum, 2009). From the Introduction: "A 2007 study by Neil Gross and Solon Simmons, two liberal academics, reported a ratio of liberal to conservative professors in social science and humanities of 9-

if they even get hired, do not dare speak up. If they do, they will not get tenure and will often lose their jobs. There is no real debate on many topics, no fresh blood, no challenge to liberal dogma. The hypocrites in academia scream about diversity but have none themselves — and diversity of thought is the most important kind of diversity.[7] When the views of half of the country are not represented, and, indeed, are deplored by most in academia (remember Hillary Clinton's "basket of deplorables"), then what comes out of academia and their accomplices in the news media — especially with regard to history — is the liberal party line, preached by liberals

1. In fields such as Anthropology and Sociology, these figures approach 30-1." http://www.discoverthenetworks.org/Articles/onepartydhjl.html, accessed January 26, 2017.

[7] **There is also rampant discrimination in hiring in academia.** People are discriminated against because of their political views. How could it be any other way when academia is overwhelmingly liberal — in some fields, as stated by Horowitz and Laksin in the previous footnote, 30 to 1 — and it has been this way for the past 50 years. Liberals discriminate against non-liberals in hiring. Liberals hire only other liberals. It is obvious that academia is a hostile work environment for everybody but liberals, and increasingly hard left liberals, because of diversity departments that demean white people, speech codes that treat conservative views as hate, anti-Christian rhetoric, etcetera, ad nauseam. This also makes much of academia extremely hypocritical — again — because in addition to screaming about diversity, which is non-existent in academia, they also scream about discrimination, yet they discriminate openly against the views of over half the country. Conservatives and other non-liberals need not apply to academia, though much of academia is funded by taxpayer money, greater than half of which comes from conservatives and non-liberals.

without fear of criticism or examination.

I know from my personal experience that many of the liberals in academia are fine people who, despite their liberal bias, try to be fair. But I know many others who are rigidly doctrinaire and definitely not fair, and they have the power structure and majority to impose their will with impunity.

These doctrinaire liberals preach their views constantly by weaving them into their classes — comments, smirks, rolls of the eyes here and there — which intimidate young students and coerce them into writing things they don't believe in order to pass.

As every honest scholar knows, to understand the past, one must view the past the way the people who lived in the past viewed it. In the past, things were almost always brutal, disease-ridden and unfair. Pain and death were always present. As English philosopher Thomas Hobbes wrote in *Leviathan*,[8] there was "continual fear, and danger of violent death: and the life of man, solitary, poor, nasty, brutish and short." In most of the past, people did the best they could to survive and get ahead in a harsh world. The world of the past was not today's middle class America but that is the standard ignorant liberals want you to judge it by.

David Harlan in his book, *The Degradation of American History*, says that, starting in the 1960s with the Civil Rights Movement, leftist historians began criticizing American history as elitist. They said it "focused our attention on great white men at the expense of women and minorities, that it ignored the racial and

---

[8] *Leviathan* was Thomas Hobbes most famous work. It was written in 1651.

ethnic diversity of national life, that it obscured the reality of class conflict." They wanted to expose the complicity of white men "in the violence and brutality that now seemed to be the most important truth about American history." They "feel no need to say what is good in American history."[9]

It's worse for Southern history.

Eugene D. Genovese,[10] one of America's greatest historians before his death in 2012, wrote this is 1994:

> Rarely, these days, even on Southern campuses, is it possible to acknowledge the achievements of the white people of the South. The history of the Old South is now often taught at leading universities, when

---

[9] David Harlan, *The Degradation of American History* (Chicago: University of Chicago Press, 1997), xv.

[10] Genovese was a brilliant historian as the following paragraph illustrates. It is the opening paragraph of an essay in *The Journal of Southern History*, Volume LXXX, No. 2, May, 2014 entitled "Eugene Genovese's Old South: A Review Essay" by J. William Harris: "The death of Eugene D. Genovese in September 2012 brought to a close a remarkable career. In the decades following his first published essay on Southern history, Genovese produced an outstanding body of scholarship, based on a rare combination of deep research in primary sources; a mastery of the historical literature, not only in Southern history but also in many complementary fields; a sophisticated command of methodological issues; and often sparkling prose. And Genovese's reputation reached far beyond specialists in Southern history, and even beyond the academy. In 2005 a reviewer in one magazine for a general readership called Genovese the 'Country's greatest living historian' and his *Roll, Jordan, Roll* 'the most lasting work of American historical scholarship since the Second World War.'"

> it is taught at all, as a prolonged guilt-trip, not to say a prologue to the history of Nazi Germany. . . . To speak positively about any part of this Southern tradition is to invite charges of being a racist and an apologist for slavery and segregation. We are witnessing a cultural and political atrocity.[11]

Dr. Genovese goes on to say that this cultural and political atrocity is being forced on us by "the media and an academic elite."[12]

In the 2016 presidential campaign, 96% of money donated by journalists went to liberal Democrat Hillary Clinton. Most of the news media are so biased,[13] it makes them untrustworthy and even more dishonest than academia. In campaign coverage, the fraudulent media colluded with Clinton and gave her debate questions in advance, allowed her campaign to edit stories, asked her campaign for advice and quotations they could use to attack Donald Trump, and made no effort to hide their contempt for objectivity.

---

[11] Eugene D. Genovese, *The Southern Tradition, The Achievement and Limitations of an American Conservatism* (Cambridge: Harvard University Press, 1994), Preface, xi-xii.
[12] Ibid.
[13] In numbers of journalists giving, 50 gave to Republican Donald J. Trump, while 430 gave to Clinton. That means 10% of journalists donated to Republican Trump, and 90% to Democrat Clinton. See David Levinthal and Michael Beckel article, October 27, 2016, "Journalists shower Hillary Clinton with campaign cash", https://www.publicintegrity.org/2016/10/17/20330/journalists-shower-hillary-clinton-campaign-cash, accessed January 25, 2017.

Too bad it backfired and greatly damaged the credibility of the media — perhaps beyond repair — just as political correctness has turned much of academia into a caricature to laugh at.

Over half the country now sees much of the "mainstream media" as liars where fake news is pervasive.[14] Think "hands up, don't shoot," which tore the country apart but never happened. However, it did meet racist liberal objectives to paint a black criminal as a victim, and a white person, a white cop doing his job, as the bad guy.

Angelo M. Codevilla,[15] in his excellent essay "The Rise of Political Correctness",[16] gives us a perfect parallel

---

[14] Some 69% of voters today (2017) "do not believe the news media are honest and truthful." See Media Research Center NewsBusters Staff article, November 15, 2016, "MRC/YouGov Poll: Most Voters Saw, Rejected News Media Bias." http://www.newsbusters.org/blogs/nb/nbstaff/2016/11/15/mrcyougov-poll-most-voters-saw-rejected-news-media-bias, accessed January 26, 2017.

[15] Angelo M. Codevilla: Claremont Review of Books contributor information states that "Angelo M. Codevilla is a senior fellow of the Claremont Institute and professor emeritus of International Relations at Boston University. He has been a U.S. Naval Officer, an Assistant Professor at the Grove City College and North Dakota State College, a U.S. Foreign Service Officer, and a member of President-Elect Reagan's Transition Team. He served as a U.S. Senate staff member dealing with oversight of the intelligence services, a professorial lecturer at Georgetown University and a Senior Research Fellow for the Hoover Institution at Stanford University." http://www.claremont.org/crb/contributor-list/116, accessed January 15, 2017.

[16] Angelo M. Codevilla, "The Rise of Political Correctness," in the Claremont Review of Books, posted November 8, 2016, Volume XVI, Number 4. http://www.claremont.org/crb/article/

between the loss of credibility of the American news media and the loss of credibility of the Communists in the old Soviet Union. He points out that the Communists were so distrusted that "whenever the authorities announced that the harvest had been good, the people hoarded potatoes; . . . ".

Same in America today, and that is what Donald J. Trump's victory signifies. Over half the country despises academia and the media and does not trust them. When the mainstream media, frothing at the mouth with liberal condescension and hate tried every sleazy trick in the book to defeat Trump, it reinforced to half the country that Trump was their man.

Academia has done the same dishonest thing with American history, especially Southern history.

The War Between the States is the defining event in American history. Out of a population of 33 million, 800,000 were killed and over a million wounded.[17] If the soldiers of World War II were killed at the same rate as the War Between the States, we would have lost 3,870,000 instead of 405,399; and we would have had 6,385,500 wounded instead of 670,846.

But history is so pathetic in this day and age that the

---

the-rise-of-political-correctness, accessed January 15, 2017.

[17] Rachel Coker, "Historian revises estimate of Civil War dead," published September 21, 2011, Binghampton University Research News — Insights and Innovations from Binghampton University, http://discovere.binghamton.edu/news/civilwar-3826.html, accessed July 7, 2014. These are the widely accepted death statistics of historian J. David Hacker of Binghampton University. He has determined a range of between 650,000 and 850,000 deaths. He splits the difference and uses 750,000. I believe it was on the higher end of his range so I use 800,000 in my books.

cause of this gargantuan event is not even studied. Historian Joe Gray Taylor noted that Pulitzer Prize winning historian David H. Donald "seems to have been correct when he said in 1960 that the causation of the Civil War was dead as a serious subject of historical analysis" and that "A 'Southern' point of view on the secession crisis no longer exists among professional historians."[18]

A Southern point of view certainly does exist.

For the South, 1861 was 1776 all over.

The North unquestionably did not invade the South to end slavery. This is provable beyond the shadow of a doubt, though that is exactly the view that the media and academia have forced on us since the 1960s. They either force it on us directly, or validate it by not challenging it (and if we disagree with them, we are racists and apologists for slavery and segregation as Dr. Genovese noted).[19]

The North invaded the South to preserve the Union as Abraham Lincoln said over and over and over — not end slavery. All Northern documents such as the War Aims Resolution, Corwin Amendment, Preliminary Emancipation Proclamation, et al., prove this

---

[18] Joe Gray Taylor, "The White South from Secession to Redemption," in John B. Boles and Evelyn Thomas Nolen, *Interpreting Southern History, Historiographical Essays in Honor of Sanford W. Higginbotham* (Baton Rouge: Louisiana State University Press, 1987), 162-164.

[19] The compiler's book, *Slavery Was Not the Cause of the War Between the States, The Irrefutable Argument.* (Charleston, SC: Charleston Athenaeum Press, 2014), makes a powerful argument and is thoroughly documented with 218 footnotes and 207 sources in the bibliography.

conclusively. These documents came about before the war or through the first two years of the war when the North was glad to state its true intentions, which it made crystal clear.

What came later such as the Emancipation Proclamation, which freed no slaves or few, were war measures after hundreds of thousands of people had been killed. They had nothing to do with why the North went to war in the first place. They and Lincoln were adamant that the North went to war to preserve the Union, and the reason for that is that Northern wealth and power were dependent on the Union and on the South.

Cotton was king and the most demanded commodity on the planet and the South had 100% control of it. Without the ability to ship Southern cotton — which alone had been 60% of U.S. exports in 1860 — and manufacture for its huge rich captive Southern manufacturing market, the North was dead. It faced economic annihilation leading straight to anarchy. Manufacturing for the South was the majority of Northern manufacturing, while shipping cotton and other Southern commodities was the majority of Northern shipping. No country can lose the majority of its manufacturing and shipping overnight without a complete collapse into anarchy.

Abraham Lincoln knew that with European recognition and military treaties, the North would not be able to beat the South militarily. The way would then be clear for the South with total control of King Cotton, to ascend to dominance in North America and the world.

These were extremely weighty issues for Abraham Lincoln, president of the North, because the entire future of the North for all time was dependent on them. He was looking at a complete shift of national power from North to South, and it was happening with lightning speed.

Going to war, however, was not a difficult decision for Lincoln.

War would solve the enormous political problems he had at that time, and it would solve his impending economic disaster. He knew, at that point in history, that the North had four times the white population of the South, most of the country's manufacturing including perhaps over 200 times more weapon manufacturing than the South, a standing army, a navy with fleets of warships, merchant shipping, a functioning government with access to unlimited immigration (around 25% of Northern soldiers ended up being immigrants), and more.

Lincoln figured he could win easily. After all, he was a 20 foot tall man loaded with modern weaponry starting a fight with a five foot tall man carrying a musket.

Of course Lincoln wanted to fight.

But what he got back was an epic amount more than he anticipated.

Henry L. Benning, one of Robert E. Lee's most able brigadier generals and for whom the sprawling U.S. Army base, Fort Benning, is named, stated before the war:

> The North cut off from Southern cotton, rice, tobacco, and other Southern products would lose three fourths of her commerce, and a very large proportion of her manufactures. And thus those great fountains of finance would sink very low.... Would the North in such a condition as

that declare war against the South?[20]

Benning's prescient analysis and the Southern view (not the cherry-picked quotations about slavery) is not studied because political correctness in academia and the news media prevent a serious study of Southern history — really American history — in this day and age, as David H. Donald stated, though it would certainly benefit students and the public to know the Southern view.

Think about the silliness surrounding Thomas Jefferson, founder of the University of Virginia and author of one of the greatest documents in the history of mankind, our Declaration of Independence. In 2016, a UVA professor — a professor! — drafted a letter and got 469 signatures of students and other professors protesting the use of quotations of Thomas Jefferson by UVA President Teresa Sullivan because Jefferson owned slaves. UVA faculty circulated the letter,[21] thus impressing

---

[20] Henry L. Benning, "Henry L. Benning's Secessionist Speech, Monday Evening, November 19," delivered in Milledgeville, Georgia, November 19, 1860, in William W. Freehling and Craig M. Simpson, *Secession Debated, Georgia's Showdown in 1860* (New York: Oxford University Press, 1992), 132. Benning was a justice on the Georgia Supreme Court before the war. Fort Benning is near Columbus, Georgia.

[21] "President of university founded by Jefferson asked to not quote Jefferson," November 14, 2016, FoxNews.com, http://www.foxnews.com/us/2016/11/14/president-university-founded-by-jefferson-asked-to-not-quote-jefferson.html, accessed November 20, 2016. UVA President Teresa Sullivan's response included: "Quoting Jefferson (or any historical figure) does not imply an endorsement of all the social structures and beliefs of his time." The following correction was posted on *The Cavalier Daily* website under an

young students that they too should hate Thomas Jefferson and, by extension, America's founding.

Can you imagine anything as shallow as a university faculty circulating a petition protesting the use of quotations of Thomas Jefferson, as towering a figure as he is in American history, because he owned slaves during a time when slavery — as horrible as it was — was legal everywhere, widespread and even many blacks in the South owned slaves?

It is as if academia wants students to be stupid, uninformed and incapable of thinking for themselves, i.e., easily led.

Academia is more interested in producing good liberal voters by intimidation and indoctrination. Many in academia don't even want conservative speakers to show up on campus and if they do, they must come with "trigger warnings" that taint their message before they utter a word. However, if any of their fragile students accidentally hear a conservative idea, there are safe spaces to run to with milk and cookies, and Play-Doh (I liked plain old modeling clay when I was in kindergarten).

Dr. Clyde Wilson, Emeritus Distinguished Professor of History of the University of South Carolina, points out that the "vast literature in recent years that has fought heatedly over Jefferson's racial views and sex life has

---

article entitled "Professors ask Sullivan to stop quoting Jefferson, Faculty, students believe Jefferson shouldn't be included in emails": "This article previously stated that student groups on Grounds collaborated to write this letter. While students and student groups signed the letter, it was drafted and circulated by University faculty." http://www.cavalierdaily.com/article/2016/11/professors-ask-sullivan-to-stop-quoting-jefferson, accessed January 19, 2017.

been carried on in an atmosphere of complete unreality."[22] Thomas Jefferson, author of the Declaration of Independence, which for the first time in human history asserted the rights of people over the rights of kings and governments (which troubles many liberals greatly), advanced the good of mankind in a gargantuan way.

Jefferson was profoundly influenced by John Locke,[23] the Age of Enlightenment's most influential philosopher. Locke's *Two Treatises on Government* discuss his revolutionary concepts of the natural rights of man, and the social contract.

The social contract is an understanding, a contract between the people and their government, meaning that the government is to protect the people and their property, and if it doesn't, it can be replaced by the people.

This is the fundamental assertion of the Declaration of Independence of 1776, and the South's secession from the Union in 1860-61. The most widely quoted phrase in the secession debate in the South in the year before Southern states began seceding comes from the

---

[22] Clyde N. Wilson, "American Historians and Their History" in *Defending Dixie, Essays in Southern History and Culture* (Columbia, SC: The Foundation for American Education, 2006), 8.

[23] John Locke is known as the father of classical liberalism, which underpins Western political thought. Classical liberalism, with its emphasis on civil liberties, rule of law and free market capitalism, is not to be confused with the fascist political liberalism of the American Democrat Party today (2017 and 50 years before), which is anti-free speech, "politically correct," and often violent.

Declaration of Independence and Locke's social contract:

> Governments are instituted among Men, deriving their just powers from the consent of the governed, That whenever any Form of Government becomes destructive of these ends, it is the Right of the People to alter or to abolish it, and to institute new Government, laying its foundation on such principles and organizing its powers in such form, as to them shall seem most likely to effect their Safety and Happiness.

How pathetic and unenlightened for the faculty of any university,[24] but especially the one founded by Thomas Jefferson, to want to forbid his quotations because he owned slaves. Intelligent people can be appalled at slavery but understand that in our evolution

---

[24] It surprised me greatly that liberals got upset that the Russians might have influenced our election, since so many liberals, especially in academia, are Marxists who adored the old Soviet Union and Communism before President Ronald Reagan defeated them both. Seems like liberals would have appreciated the Russian influence. The Russians and Wikileaks, in the 2016 presidential campaign, did the job of our bigoted, incompetent news media, and exposed extreme media collusion with the Clinton campaign and Democrat Party such as a CNN reporter and head of the DNC, Donna Brazile, who gave debate questions to Clinton in advance (and Clinton gladly accepted them), another "journalist" who let the Clinton campaign edit stories, another who asked the Clinton campaign for things he could use to bash Donald Trump, and another who stated clearly that they should not be objective but should be the opposition party to Trump. And three-fourths of them were, and are, as of this writing (2017).

as a nation, slavery existed for a while,[25] as with most nations on earth, though slavery has been gone for a century-and-a-half.

Dr. Wilson states that this nonsense about Jefferson

> proceeds on the assumption that Jefferson was essentially a twentieth century middle class American rather than an eighteenth-century Virginia planter. This is not simply the common mistake of reading the present into the past. It is a pervasive intellectual confusion that runs unchecked and unrecognized through both our popular and academic history.[26]

Dr. Wilson observes that "The main theme of American history is being shifted from national unity and national achievement". The "transformation of American history from an account of the building of a new nationality to the celebration of an ethnic collage is not a result of the discovery of new knowledge."[27] It is "the actual destruction or suppression of old views, and their replacement by others newly manufactured for social

---

[25] New Englanders and the British before them brought most of the slaves here and made huge profits in the process. Slave-picked cotton made the North rich and powerful. Slavery was not expanding in 1860 but contracting, and the slave trade had been outlawed for 52 years in 1860. The industrial revolution with great new labor-saving farm machinery would have killed slavery with nobody dying, and no excessive hate.
[26] Wilson, "American Historians and Their History" in *Defending Dixie, Essays in Southern History and Culture*, 8
[27] Ibid., 10.

purposes rather than as a consequence of knowledge."[28]

Sounds like what Orwell warned us about in *1984* when Winston Smith lamented

> Do you realize that the past, starting from yesterday, has been actually abolished? If it survives anywhere, it's in a few solid objects with no words attached to them, like that lump of glass there. Already we know almost literally nothing about the Revolution and the years before the Revolution. Every record has been destroyed or falsified, every book has been rewritten, every picture has been repainted, every statue and street and building has been repainted, every statue and street and building has been renamed, every date has been altered. And that process is continuing day by day and minute by minute. History has stopped. Nothing exists except an endless present in which the Party is always right. I *know*, of course, that the past is falsified, but it would never be possible for me to prove it, even when I did the falsifications myself. After the thing is done, no evidence ever remains.[29]

Falsification of the record is the essence of political

---

[28] Ibid., 5.

[29] George Orwell, *1984* (New York: New American Library, 1950), 128.

correctness and it is "atrocious treason" as Dr. Johnson (Samuel Johnson) writes in Rambler No. 136.

> To deliver examples to posterity, and to regulate the opinion of future times, is no slight or trivial undertaking; nor is it easy to commit more atrocious treason against the great republic of humanity, than by falsifying its records and misguiding its decrees.

Dr. Wilson goes on:

> Even when it is not badly distorted, academic history has become, not the remembered story of human life but only a commentary on dogma. . . . It converts great segments of humanity into oppressors who deserve only annihilation. The result is today's academic history — a weird combination of supposedly objective 'social science' and romantic exaltation of favored minorities designated as the oppressed. This history fails both as accurate record and as material for social comity. As Christopher Lasch pointed out years ago, scholars have abandoned the search for reality in favor of the classification of trivia. But it is worse than that. It is in the nature of dogma that dissenters are quickly suppressed. Conformity of opinion about what is

> significant and true about the past has never been as rigorous among academic historians, and all who listen to them, as it is today.[30]

Academia is able to get away with this because there is no diversity of thought or debate to challenge it. The left does not want debate as we saw February 1, 2017 at UC-Berkeley when conservative speaker Milo Yiannopoulos had to be rushed off campus when a riot erupted in which fires were set, windows smashed, a girl pepper-sprayed in the face on national TV, etc. There was only one arrest, which tells violent leftists that liberal administrators are on their side: Put on black masks and come destroy campuses, set fires, use sledge hammers, pepper spray and other weapons when conservatives speak as the thugs at UC-Berkeley did recently. Be as violent as you want because you are our brown shirt heroes and will not be prosecuted.

Liberal discrimination by academia in hiring only liberals, has given them an absolute protected liberal environment (paid for with taxpayer money from over half of the country that despises them) with which to

---

[30] Wilson, "Scratching the Fleas: American Historians and Their History" in *Defending Dixie, Essays in Southern History and Culture*, 47. Those "favored minorities" are found in the Democrat Party, which is itself defined by identity politics: race, class, gender, sexual orientation, etc. The Democrat Party does not represent Americans in the aggregate. It represents groups of Americans, thus history is being rewritten by liberal academia and promoted by liberals in the media to favor liberal Democrat groups and spew hate on everybody else, especially those who disagree with them.

impose their bigoted intolerant views.

Many in academia are cowards because they know if they run afoul of political correctness they can have their careers destroyed. Again, they know that to say anything good about the Old South in this atmosphere of hate and censorship invites the charge of being a racist and apologist for slavery and segregation as Dr. Genovese stated. They would rather tell lies and keep their paychecks coming.

Academia has been overwhelmingly liberal for a long time, with little diversity of thought and much pressure to conform, and so has the news media. Neither are going to change, but the difference today is that over 70% of the country do not take either of them seriously and indeed despise their bigotry.

About history, Dr. Wilson states that "The young person must be able to make his nation's history his own, make it a history of his own 'fathers,' just as was done, until a generation or so ago."[31] Today, however, young people are taught by academia to hate their country because they are descended from vile oppressors or the oppressed.

> Most of the work of academic historians today can portray the American story in no other terms except as an abstract fantasy of oppressors and oppressed. No society has ever had more professional historians and devoted more resources to historical work of all kinds than modern America—

---

[31] Wilson, "American Historians and Their History" in *Defending Dixie, Essays in Southern History and Culture*, 7.

> or produced so many useless, irrelevant,
> and downright pernicious products.[32]

Angelo M. Codevilla agrees that there is a revolution going on and it's "all about the oppressed classes uniting to inflict upon the oppressors the retribution that each of the oppressed yearns for" because, as liberals see it, "America was born tainted by Western Civilization's original sins — racism, sexism, greed, genocide against natives and the environment, all wrapped in religious obscurantism, and on the basis of hypocritical promises of freedom and equality."[33]

I saw a man-on-the-street interview recently with a white male college student. He said he did not vote because America was a racist nation founded on stealing land and slavery, and he hated our country.[34]

On February 13, 2017 I tuned into Fox News and Jesse Watters was interviewing a college woman who had been chanting "Two, four, six, eight, America was never great." He asked her why she was chanting that and she said because it rhymed. He pressed and said "You really don't think America is great?" to which she said no. He said, what about us defeating the Nazis? She shrugged her shoulders.

This is the essence of the political correctness and

---

[32] Wilson, "Scratching the Fleas: American Historians and Their History" in *Defending Dixie, Essays in Southern History and Culture*, 45.

[33] Angelo M. Codevilla, quotations from "The Rise of Political Correctness".

[34] This interview occurred in January or February 2017 on Fox News one afternoon. I tuned in as it was going on so do not know the context.

hate-America liberal indoctrination she is getting in the classroom, on campus, and in much of the news media. This indoctrination is also illustrated well by the attacks on Thomas Jefferson by the pathetic UVA faculty.

American history should be an inspiring and inclusive story of our country. The darker parts should not be whitewashed but neither should they define the whole.

The historians of the past are extremely important today. When a historian such as Ramsdell writes about the Fort Sumter incident and beginning of the war, our cultural standards today are irrelevant to that argument.

Ramsdel's treatises in this book can be considered primary sources themselves of a sort. They demonstrate the state of historiography up to the early 1940s when Ramsdell died.

I agree with every word in his two most famous treatises: "Lincoln and Fort Sumter" and "The Natural Limits of Slavery Expansion". Both are as powerful and apropos today as the day they were written.

Ramsdell proves, in "Lincoln and Fort Sumter," that Abraham Lincoln engineered the beginning of the war in Charleston Harbor when he sent a hostile naval expedition loaded with artillery troops and ammunition into the most tense situation in American history. It was his intent to start the war as many Northern newspapers admitted. The *Providence (R.I.) Daily Post* wrote, in an editorial entitled "WHY?", April 13, 1861, the day after the commencement of the bombardment of Fort Sumter:

> We are to have Civil War, if at all, because

> Abraham Lincoln loves a party better than he loves his country. . . . Mr. Lincoln saw an opportunity to inaugurate civil war without appearing in the character of an aggressor.

From his standpoint, Lincoln had to get the war started as fast as he possibly could. There was no reason whatsoever for him to wait. With every second that went by, the South got stronger and the North got weaker. His economy was heading fast into complete annihilation and the moment Confederates established trade and military alliances with Great Britain and Europe, the North would not be able to beat the South. The South, with 100% control of the most demanded commodity on the planet — cotton — would then ascend to dominance in North America and the world.

Ramsdell ends "Lincoln and Fort Sumter" with absolute proof that Lincoln started the War Between the States: the diary entry of Lincoln's good friend, Orville H. Browning, in which Browning recorded Lincoln's exact words, as told to him the night of July 3, 1861 by the usually closed-mouth Lincoln. Abraham Lincoln bragged about deliberately starting a war that ended up killing 800,000 Americans and wounding over a million, to save himself and the Republican Party politically.

Ramsdell's "The Natural Limits of Slavery Expansion" proves that slavery was not extending into the West. One prominent historian called the slavery in the West issue a bogus issue about an "imaginary Negro" in an impossible place. Two of the Western territories had been open for slavery for 10 years and there were only 24 slaves in one, and 29 in the other. Slavery only worked on rich cotton soil near rivers or railways on which cotton could be transported.

Within 20 years of the end of the war, slavery would have ended in the United States by the industrial revolution and technological advancements in farm machinery that would pick the cotton much faster than slaves and at a fraction of the cost. Ramsdell, and an increasing number of historians today, maintain that the War Between the States was a totally unnecessary war, and I agree.

Ramsdell's treatises "General Robert E. Lee's Horse Supply, 1862-1865," "The Confederate Government and the Railroads" and "The Control of Manufacturing by the Confederate Government" are the most enlightening I have ever read as to why the South won the early part of the war, but wore down due to massive Northern industrial and other resources.

Lee's horses, after 1862, were often half-starved, sick, impossible to replace thus his cavalry was severely restricted on the battlefield, but also his artillery because it took horses to pull the cannons and other ordnance. Ramsdell writes this toward the end of "General Robert E. Lee's Horse Supply, 1862-1865":

> With his flank turned and his remaining communications about to be cut, Lee began at once the withdrawal which he had long foreseen must be made. It would have been a difficult operation with his animals in good condition; but now at the end of a severe winter when they were weak and

> slow from exposure and starvation it was a desperate undertaking. Only the stronger teams were able to take out wagon trains and guns, and on the forced marches without food they soon broke down. The cavalry could not keep pace with the better horses of Sheridan. At the end of a week what was left of a proud army was surrounded and the long struggle was over.

The problems with critical rail transportation were just as dire. Ramsdell writes in "The Confederate Government and the Railroads":

> For more than a year before the end came the railroads were in such a wretched condition that a complete breakdown seemed always imminent. As the tracks wore out on the main lines they were replenished by despoiling the branch lines; but while the expedient of feeding the weak roads to the more important afforded the latter some temporary sustenance, it seriously weakened the armies, since it steadily reduced the area from which supplies could be drawn.

All of the other treatises are extremely enlightening too. You can tell the view with which historians of the past looked at history and how it affected their interpretations. Even with the perspective of a different time, the vast majority of the history of Ramsdell and his colleagues is solid as a rock — and in fact includes much important information long overlooked or discounted by the politically correct frauds of today.

Ramsdell's book reviews are works of art. He reviewed many of the books we still hold in high esteem such as *R. E. Lee: A Biography*, by Douglas Southhall Freeman; *Life and Labor in the Old South*, by Ulrich Bonnell Phillips; *The Civil War and Reconstruction*, by J. G. Randall and 12 others (which are just a handful of Ramsdell's reviews). Included are reviews of books by famous historians such as Frederick Jackson Turner, creator of the Frontier Thesis.

This book — *Charles W. Ramsdell, Dean of Southern Historians, Volume One: His Best Work* — is an important book; and, *Volume Two: His Texas Treatises*, will be out by the end of summer, 2017 followed fast by a third book centered around Ramsdell's "Lincoln and Fort Sumter".

The treatises, book reviews and citation are all verbatim as they appeared originally. Nothing has been edited out or added except for some additional explanatory footnotes. The footnotes have all been renumbered to run continuously throughout the book.

Most of the punctuation and capitalization in the treatises, book reviews and notes are exactly as written by Ramsdell and edited by the various scholarly publications in which they appeared. Some of it is not as we would do today but it doesn't matter one iota. There is nothing that is not understandable in any of it. It is just different here and there, and I wanted to acknowledge that.

It is nice to have mostly Confederate names for battles such as Manassas for Bull Run, and Sharpsburg for Antietam.

As stated, I am very proud to bring out the writings of Charles W. Ramsdell, Dean of Southern Historians, and others who were brilliant and uncompromised by political correctness. There is MUCH more to come.

Gene Kizer, Jr.
Charleston, South Carolina
April 12, 2017

Charles W. Ramsdell

Dean of Southern Historians

Part One

The Dean

Charles W. Ramsdell: Historian of the Confederacy[35]

by

Wendell Holmes Stephenson

"My God! Has it come to that?"

More than a score of years ago, Charles W. Ramsdell was invited to inaugurate the Walter Lynwood Fleming Lectures in Southern History at Louisiana State University. A program of the first series, characterizing the lecturer as the "Dean of Southern Historians," reached him a few days before he left Austin for Baton Rouge. The well-merited designation brought a quick response from the historian who was modesty personified: "When I read that I could not refrain from saying to myself, 'My God! Has it come to that?'"[36]

It had, indeed, though it would be difficult to assign a mono-cause for the respect and esteem in which Ramsdell was held by the historical guild in the 1930's. He had, by that time, served as president of the Mississippi Valley Historical Association and of the newly established Southern Historical Association.[37] Faithful

---

[35] Wendell Holmes Stephenson, "Charles W. Ramsdell: Historian of the Confederacy", *The Journal of Southern History*, Vol. 26, No. 4 (Nov., 1960), 501-525. Article and citation are verbatim.

[36] Charles W. Ramsdell to the writer, April 11, 1937, in writer's personal files.

[37] In addition to the presidencies of the two associations, Ramsdell gave other official services to professional societies:

and competent member of the history faculty at the University of Texas since 1906, he had taught annual generations of undergraduate and graduate students for thirty years, and he had given generously of his time to directing masters' theses and doctoral dissertations. His publication record-in pages he had put in print-was certainly not impressive. His only scholarly book at that time, *Reconstruction in Texas,* was his dissertation at Columbia, printed in 1910. A score of articles and essays had appeared in such professional periodicals as the *Southwestern Historical Quarterly,* the *Mississippi Valley Historical Review,* the *Journal of Southern History,* and the *American Historical Review.* Half a hundred appraisals of books had been contributed to scholarly journals. The *Dictionary of American Biography* had profited from his concise sketches.[38]

---

secretary-treasurer of the Texas State Historical Association, 1907-1942; associate editor of the *Southwestern Historical Quarterly,* 1910-1938; member, Executive Council, Southern Historical Association, 1935-1939; member, Board of Editors, *Journal of Southern History,* 1937-1940; member, Executive Committee, Mississippi Valley Historical Association, 1928-1935; member, Board of Editors, *Mississippi Valley Historical Review,* 1930-1933; member, Executive Council, American Historical Association, 1931-1934.

[38] For "A Bibliography of the Writings of Charles W. Ramsdell," see Charles W. Ramsdell, *Behind the Lines in the Southern Confederacy,* edited with a foreword by Wendell H. Stephenson (Baton Rouge, 1944), 123-36 [This bibliography is reprinted in Part Five of this book with additional material]. Two other books had appeared by 1937: *A School History of Texas* in collaboration with Eugene C. Barker and Charles S. Potts (Chicago, 1912) and *The History of Bell County* (San Antonio, 1936), under Ramsdell's editorship. Two more were published subsequently: *Laws and Joint Resolutions of the*

Perhaps a few score of Ramsdell's Southern contemporaries equaled his productive record, and many surpassed it.

The "deanship," nevertheless, was hardly a specious invention. Contemporaries in the guild, including his students, recognized the superior qualities of his mind, the logic of his thought, the soundness of his judgment. A younger member of the craft who listened spellbound as Ramsdell read his Mississippi Valley Historical Association presidential address, "The Natural Limits of Slavery Expansion," imagined the dignified Texan attired in judicial ermine. The neophyte was so awed that he could not muster the courage, after adjournment, to make his presence known to the Association's retiring president. Five years would elapse before the anonymous admirer meekly sought Ramsdell's aid in organizing the Southern Historical Association and in launching its *Journal of Southern History.* To his amazement he discovered that the Texan was one of the most approachable men he had ever known. The self-depreciating scholar stood in sharp contrast with inhabitants of the historical world who wore their scholarship so conspicuously on their coat sleeves. He was very human, aware of his own limitations, eager to grasp further insight.

The year 1877 was significant in Southern history as the end of Radical Reconstruction and in Southern

---

Last Session of the Confederate Congress (November 7, 1864-March 18, 1865), Together with the Secret Acts of Previous Congresses (Durham, 1941) and his Walter Lynwood Fleming Lectures in Southern History, *Behind the Lines in the Southern Confederacy.* A few articles and reviews also appeared after 1937.

historiography as the beginning of Charles William Ramsdell. His birthplace was Salado in Bell County, Texas, where he attended the public schools and also the private Thomas Arnold High School.[39] The "classical course" at the University of Texas, where he received the bachelor of arts degree in 1903, included among other subjects Greek, Latin, German, and English, as well as history and political science. In most of his courses Ramsdell earned grades of A or B notwithstanding participation in football.[40] Collegiate experience on the gridiron later enhanced his value as a faculty member at his alma mater, for he understood "men and how to handle them," and "proved himself a very useful man on committees concerned with student activities."[41] Athletes were not subsidized in Ramsdell's undergraduate days. In the summer preceding his senior year, he wrote to George P. Garrison, head of the history department, indicating the need for employment to continue his quest for a degree. "Would the work that I might get on the *Quarterly* [of the Texas State Historical Association] amount to very much?" he inquired. While he preferred quarters where he could earn his board, he would labor at any task.[42] A fifth year at the University yielded the master's degree in 1904, with a thesis on Presidential

---

[39] [Charles W. Ramsdell] to the University of North Carolina Press, October 23, 1933, in Charles W. Ramsdell Papers (in Mrs. Ramsdell's possession, 1945).
[40] Transcript of record, University of Texas.
[41] Eugene C. Barker to Ulrich B. Phillips, April 2, 1911, in Eugene C. Barker Letters (Archives Collection, University of Texas, Austin).
[42] Ramsdell to George P. Garrison, July 18, 1902, in George P. Garrison Papers (Archives Collection, University of Texas).

Reconstruction in Texas.

An enviable academic record led to a fellowship at Columbia University for the year 1904-1905.[43] Ramsdell kept Garrison informed of his progress, the academic opportunities at the University, the caliber of his instructors as scholars and teachers, and the extracurricular advantages of living in New York City. Despite the "strenuous life" imposed by his teachers, he found time to observe the great metropolis—"this noisy quarter of the world." The student attended the best operas and theaters, listened to well-known preachers, and visited "points of greatest interest from the Ghetto to the Metropolitan Museum of Arts."[44]

These diversions consumed only a minimal slice of his time budget. He was ever busy attending lectures and preparing term reports. His professors passed in review as he confided to his Texas mentor the nature of their courses and the measure of their personalities. He studied American constitutional history with John W. Burgess, who seemed "to stand afar off," though his lectures were excellent. John Bassett Moore, who guided his study of diplomatic history, was "a very pleasant man," and he liked him immensely. Herbert Levi Osgood "would be a fine man to know if one could get close to him." He was "above all things a *driver,* a worker." Ramsdell thought he "would have been a great success as a restless, bustling business man. The remorseless energy with which he can pile up millions of facts and hurl them at you is overwhelming." Like his volumes on *The American*

---

[43] *Columbia University Catalogue and General Announcement, 1904-1905* (New York, 1905), 495.

[44] Ramsdell to Garrison, April 17, 1905, in Garrison Papers.

Colonies, lectures were "wearisome in their particularity."[45] Students dreaded his rapidly delivered "paralyzing" performances. But they had "an immense respect for his scholarship," and they understood why he acquired recognition as "the first authority in Colonial History." Ramsdell's thesis-length term report, a history of the Committees of Correspondence, 1764-1775, required more than a month, but he had his "reward in that Prof. Osgood seemed pleased and—which is not his habit at all—spoke of it in a rather complimentary way."

The Texan's favorites were James Harvey Robinson and William A. Dunning, both of them "charming men, not only scholarly and fine teachers, but they are very *approachable* and one can get a great deal from the personality of the men." The student was "particularly impressed" with Robinson and his course on the cultural history of Western Europe. It was "the only thing of its kind offered in the country," Ramsdell reported. Members of the class acquired firsthand acquaintance "with every writer with any claim to prominence from Boethius to Erasmus." A new approach to the French Revolution reserved a semester for the old regime, for Robinson was convinced "that the real revolution was practically accomplished by 1789." Ramsdell's forty-page report on The Republican Spirit in France before 1792 went slower than papers in other courses, as much of the evidence was in French. A formal statement of Robinson's "New History" would not appear until after Ramsdell left Columbia, but the inspiring teacher gave the student a

---

[45] See account, from which the quoted words are taken, in Dixon Ryan Fox, *Herbert Levi Osgood, an American Scholar* (New York, 1924), 47, 146-49.

genuine love for modern European history and a desire to inaugurate work in that subject when he was offered an instructorship at the University of Texas.

As a master's candidate, Ramsdell had, as we have seen, investigated Presidential Reconstruction in his native state, and he therefore felt much at home in Dunning's seminar. His report on "the legislative and constitutional history of the Civil Rights Bill" required another four or five weeks. Ramsdell was grateful for Dunning's "expression of pleasure in the paper." He was also indebted to his director for a part-time teaching position at Barnard College. As fellowships at Columbia were not renewable for a second year, he faced the problem of continuing there until residence requirements were completed. It was Dunning, the student thought, who was chiefly responsible for providing the means—an assistantship created by the prolonged absence of James T. Shotwell in Europe. As there were other candidates for the position, Ramsdell was hard pressed to find a reason for his own selection. "It all means I suppose that Prof. Dunning wants that thesis on Reconstruction in Texas . . . finished." He would teach Barnard classes in general history and while the remuneration was not great, he would acquire valuable teaching experience, a "closer contact with the faculty," and an opportunity to observe "the inside workings of a great institution"; or, as Dunning whimsically phrased it, "'in being a cog in the machine—*once in a while*.'" Ramsdell was so excited over the appointment that he required over a week for complete recovery of equilibrium.[46]

---

[46] 11Ramsdell to Garrison, December 11, 1904; April 17, 1905, in Garrison Papers.

In his mature years, the former assistant of 1905-1906 humorously recalled his first class of girls at Barnard College. He labored assiduously in preparing lectures in advance, and worked out material for at least three. The notes moved more rapidly than he anticipated, and he used all of them in the first half hour of his opening lecture. At this embarrassing point he decided that a summary of what he had said was in order. Time was still heavy on his hands, so he summarized the summary. As he had no heart for summarizing the summary's summary, he dismissed the class five minutes early.

Before Ramsdell's second year at Columbia was well under way, glorious news arrived from Austin: a letter from Garrison offering him an instructorship in history at his alma mater. As the ambition of every transplanted Texan seems to be a return to Texas, the offer—though not the salary—was highly gratifying. He thought of all the reasons why he should receive $1,200 for fifteen or eighteen hours of teaching; but as that salary was not forthcoming, he accepted the position anyway.[47]

Thus began Ramsdell's career as historian at the University of Texas, where he would continue until his death in 1942. The department of history then consisted of four members: Garrison, who was nearing the end of his period of service; Eugene C. Barker, who would continue at the University until his retirement nearly a half century later; Herbert E. Bolton, destined for a conspicuous career at the University of California after a period of seasoning; and Ramsdell, "the lowest in rank of

---

[47] Ramsdell to Garrison, November 12, December 8, 19, 1905; January 24, February 18, April 4, 1906, *ibid.*

the four men in the department," as the new appointee phrased it. Before Barker succeeded Garrison as chairman of the department a few years later, the neophyte taught ancient history, modern European history, and English history. In the last field he soon developed British Empire and period courses for advanced undergraduates and graduate students. The department's personnel expanded rapidly under Barker's leadership, and Ramsdell gradually shifted to United States history, Civil War and Reconstruction, and the Old South.[48]

Through no fault of his own, advancement came slowly in Ramsdell's early years at Texas. Despite his experience at Barnard College, the Texas authorities "considered him as in a measure untried." A new president with conservative notions of promotions and a governor who "pledged the Regents" to make no salary increases in return for a new library appropriation delayed his progress. And so the matter stood when Ulrich B. Phillips, another Dunning student, resigned his professorship at Tulane University in 1911 and sought to make Ramsdell his successor. But somehow Tulane's President Craighead got the impression that the Texan was "a factionalist and of malcontent disposition."[49] This misconception prompted a letter from Barker, full of honest if restrained commendation. "It would be hard to imagine a man further removed from the character of a factionalist than is Ramsdell," he wrote. He was "a

[48] *Bulletin of the University of Texas . . . Catalogue* (title varies), 1907-1941 (Austin), *passim.*

[49] Ulrich B. Phillips to Barker, March 30, 1911, in Barker Letters.

conservative progressive" who talked little, and least of all "about his own grievances." Ramsdell was "a good teacher, not striking, but quite well up with the average, and perhaps a little better." As a man he was "plain, straight-forward, and likeable." If he lacked "brilliant qualities," he was nonetheless "perfectly solid." Barker hoped that Texas would be able to retain his services.[50] After five years as instructor, he was promoted to adjunct professor of American history.

In evaluating Ramsdell as a historian of Reconstruction, it is necessary to understand, first, that his only serious venture into this period was his dissertation at Columbia; and, second, that it conformed rather rigidly to the generally accepted pattern of the time. As a product of the Dunning School, it was an adequate performance, though the doctoral candidate did not attain the broad perspective achieved by his contemporary, Walter L. Fleming, whose 800-page *Civil War and Reconstruction in Alabama* treated both controversial epochs on a substantial antebellum base, and supplemented conventional approaches by incorporating a great mass of social and economic data in his narrative. Ramsdell's briefer *Reconstruction in Texas*

---

[50] Barker to Phillips, March 21, April 2, 1911, *ibid.* When Dunning learned that Ramsdell might not be promoted in 1911, he wrote to a friend at Georgetown College in Texas: "If this is the case I think it is one of the greatest outrages, and worse than that; the greatest blunders ever made by an educational institution. I have the highest possible respect for Ramsdell as a man and a scholar. . . . If there were a position open at Columbia we of the History Department would offer it to him in a minute, so sure are we of his powers to do good work." William A. Dunning to S. H. Moore, March 9, 1911, *ibid.*

devoted only sixteen pages to secession and the course of the Civil War in Texas. Primarily political and constitutional history, it provided only incidental consideration of social and economic aspects of the period. The brief presentation of labor conditions was certainly inadequate, and the state's history out of social context made the study less than comprehensive. The two-page bibliography indicated chief reliance upon Texas newspapers, convention and legislative journals, and correspondence of contemporaries.

Judging Ramsdell's performance in the conceptual framework of the Dunning School, it was on the whole an impartial effort at factual reporting. The historian frankly admitted, however, that he was "naturally drawn into a sympathetic attitude toward the people whose social and political system was being 'reconstructed.'"[51] Avoiding injudicious terminology throughout most of the book, Ramsdell's irritation occasionally came to the surface. He spared no words in excoriating the Radicals. "Never, perhaps, was punitive legislation founded upon a more distorted array of evidence, upon worse misrepresentation as to facts." Governor Edmund J. Davis was "self-willed, obstinate, pig- headed almost beyond belief, a most intense and narrow partisan." Ramsdell credited him with personal honesty, but he was associated with "a group of the most unprincipled adventurers that ever disgraced a government." The Radical governor's police included "the worst desperadoes in the state." With the badge of public authority, they "committed the most high-handed

---

[51] Charles W. Ramsdell, *Reconstruction* in *Texas* (New York, 1910), 8.

outrages: bare-faced robbery, arbitrary assessments upon helpless communities, unauthorized arrests, and even the foulest murders." Lawlessness was not confined to the Radicals, however, for "Texas was pre-eminently a frontier state," with rough and tumble means of settling disputes outside the law. The state's problem "was to endure as best she could the rule of a minority, the most ignorant and incapable of her population under the domination of reckless leaders, until time should overthrow it. Reconstruction had left the pyramid upon its apex; it must be placed upon its base again."[52] Whether in the period before Radical control began, or subsequent to it, Ramsdell defended Conservatives against charges of disloyalty and approved their means of restoring "home rule."[53]

Aside from the emphasis on political and constitutional history to the neglect of social and economic factors, the study's major imperfections lay, first, in the writer's inability to detach himself from the period's controversial events and personalities; and, second, in his acceptance of the contemporary view that the Negro was innately inferior. Southern writers, and Northern historians too, had not yet taken the Negro seriously as a factor worthy of careful consideration. Another generation would pass before his part in Reconstruction would be fairly evaluated, or progressive and constructive implications of the period would be recognized and appreciated. Ramsdell understood the period only in part. Understanding eventually became a quality of his mind, but it would be directed mainly to the

---

[52] *Ibid.*, 148, 317, 302, 127, 292.
[53] *Ibid.*, 66-67.

Civil War period and its antebellum background.

Before a history of the Confederacy could be written, its records must be assembled for the historian's use. The initial impulse at the University of Texas was supplied by Barker, with Ramsdell an enthusiastic coadjutor. The unfair accounts that came out of the North were not due to prejudice alone. The South was responsible in large part for the unbalanced history that was written, for the region had done little to make its records available. Barker's 1912 questionnaire to leading historians and librarians in the South provided a dismal picture of meager funds for acquisitions and inevitable paucity of materials for the teaching of Southern history and slight incentive for writing it.[54]

The origin of the Littlefield Fund for Southern History needs no extensive treatment here, for it is a well-known story. Tactful persuasion led a Confederate veteran, Major George W. Littlefield, to donate an endowment of $25,000 in 1914 for the purchase of Southern records, to provide special gifts totaling over $30,000 by 1920, and to bequeath an additional $100,000 to the endowment at his death in that year. From the early 1920's the Fund yielded an annual income of about $4,000. Ramsdell became a member of the Littlefield Committee in 1914 and negotiated for the acquisition of sundry important purchases from the beginning of his membership. He was immediately in charge of two of the three projects launched in 1937. One

---

[54] Returned copies of the questionnaire and correspondence relating to it are in the Barker Letters; transcripts in possession of the writer. See also Barker to Editor, May 1, 1914, New York *Nation,* July 2, 1914.

of them was a program of microfilming Southern materials for the three score years from 1820 to 1880; the other was a proposal for writing a multivolume history of the South.[55]

As director of the microfilm project, Ramsdell arranged for a graduate student, Barnes F. Lathrop, and his wife, to tour the important depositories from Massachusetts to Louisiana, equipped with a Photorecord camera. For two years, 1937-1939, and again in the summer of 1940, they photographed nearly forty thousand feet of film, representing approximately four hundred thousand pages of historical material. "The Ramsdell films," Lathrop reported, "constitute one of the most valuable additions to the Littlefield collection."[56]

The other Littlefield Fund project was a history of the South, for the endowment contemplated the writing of the

---

[55] [Barker] to Major George W. Littlefield, April 16, 1914; April 6, 1915; "about April," July 18, 1917; Littlefield to Barker, October 16, December 2, 1914; April 7, 1915; March 5, 1917; [Barker] to Clarence Ousley, chairman, and the Board of Regents, April 11, 1914; editorial in the *Texan,* May, 1914; [Barker] to F. W. Cook, chairman, Board of Regents, April 26, '1915; ['Barker] to President W. J. Battle, April 21, 1916; [Barker] to President R. E. Vinson, April 23, 1917; J. E. Goodwin, Librarian's Report on Expenditures for the Littlefield Collection for Southern History, all in Barker Letters. See also Paul W. Schroeder, "The Littlefield Fund for Southern History, I. A History of the Littlefield Fund," *Library Chronicle of the University of Texas,* VI, No. 1 (Spring 1957), 3-23; Thomas F. Harwood, "The Littlefield Fund for Southern History, II. Catalogued Books and Pamphlets on the Negro, ,Slavery, and the Civil War," *ibid.,* No. '2 (Spring 1958), 3-16.

[56] Barnes F. Lathrop, "Microfilming Materials for Southern History," *Journal of Documentary Reproduction,* II (June 1939), 91-108.

South's history as well as the assembling of its records. The Littlefield Committee delegated Ramsdell to plan a series of volumes. At first he contemplated three, covering the period from the end of the Revolution to the close of Reconstruction. He later decided to add three more to include the earlier and later periods of Southern history. It does not detract from the Texan's initial planning to understand that a somewhat more ambitious project was in an advanced stage at a neighboring university, and that its ten-volume proposal was eventually adopted as a joint endeavor of the Littlefield Fund and the Louisiana State University Press. Ramsdell's capacity for adjustment and co-operation came to the surface in working out arrangements for a history that involved co-operation between the sponsoring institutions, between the co-editors, and among the contributors of the several volumes. The editorial correspondence is replete with evidence of Ramsdell's sound judgment in choosing authors and in establishing flexible objectives for the whole series.[57] Tragedy intervened to prevent him from writing the Confederacy volume or from witnessing the appearance of even the first published number, which materialized five years after his death. The planning stage of the co-operative work yields tangible evidence of Ramsdell's philosophy of history.

As anyone who knew the workings of Ramsdell's mind might expect, he was opposed to system builders—to great patterns or unifying themes. Using the history of the South as an example, he explained his view in unmistakable terms. As he contemplated the broad

---

[57] Correspondence of Ramsdell and the writer, in writer's personal files.

range of the South's past, he saw

> a continuous evolution (mixed up with a certain amount of *devolution*), but it is too complex for my simple mind to reduce to a few simple factors. I am conscious of continuous change and a multiplicity of factors some of which are simple and evident, some of which I only partly understand, and others that I merely sense without being able to analyze them. Some are local, some are world-wide. There are intangibles that elude me, imponderables that I cannot weigh.

Ramsdell hoped that each contributor to the co-operative work would

> avoid treating his volume as a mere cross-section of the South's history, but rather as a chapter in a continuous narrative; he should look both backward and forward in order that, while telling his portion of the story as completely and effectively as possible within the limits set for him, he may fit it into the larger unit.

With Phillips' "most interesting paper" on "The Central Theme of Southern History" Ramsdell was "favorably impressed"; it was, he thought, "factually consistent" and "as valid as any." But he saw "an element of danger" in such a fascinating "intellectual exercise." "It could so easily be pushed too far."[58] Writing to Avery Craven, who requested Ramsdell's view of the article, the Texan pointed out the difficulty,

> considering the complexity of conditions, interests and motives throughout the whole South during its whole history to select any one thing as *the* central theme. But if we must have one, I think the problem of racial adjustment or white control comes nearest being the one of universal interest.[59]

The amiable Ramsdell did not permit his charitable nature to dull the edge of critical appraisal. In reviews and in correspondence he defended Clio against assaults of historians who distorted facts, drew erroneous conclusions, or wrote unbalanced history. He could speak most authoritatively on the Confederacy period, and with some assurance on the Old South and the Civil War's aftermath. Phillips' primacy in the ante bellum field did not overawe him. *Life and Labor in the Old South* was a "delightful and illuminating book," and it approached definitiveness in the author's main interest—"southern agriculture and agricultural society." But he protested neglect of the six million non-slaveholders and scant attention to business, industry, transportation, and education. Specialists in Southern history had a right to expect answers to many questions which the book did not

---

[58] Ramsdell to the writer, June 16, 1939, *ibid.*

[59] Ramsdell to Avery 0. Craven, October 29, 1928, in Ramsdell Papers.

provide. Was the peculiar institution profitable? Did it actually oppress small farmers? "Barring the cataclysm of war," what was "the probable future of the institution?" What had agricultural reformers accomplished by 1860? And was the time propitious for a "fairly workable synthesis" of the Old South?[60]

Ramsdell's most caustic criticism was reserved for Frank L. Owsley's *State Rights in the Confederacy.* He disagreed with the book's thesis, that the Confederacy "Died of State Rights." The author did, of course, "find evidence that state authorities frequently hampered the Confederate government," and therein lay a "useful contribution." But Owsley "tried to prove too much," and he did so by mishandling evidence. Becoming enamored of his subject, he abandoned "his critical powers," he "accepted isolated and casual statements as bases for sweeping declarations," he "read into some of his sources statements that are not there even by implication," and he "ignored evidence that tends to disprove or to qualify materially portions of his general thesis." Owsley assigned Southern governors-particularly Joe Brown, Zebulon B. Vance, and Francis R. Lubbock-to the "rogues' gallery," but he found no fault with any of the Confederate government's acts or policies or any of its officials' mistakes. The dissertation's thesis, Ramsdell thought, was more appropriate for the period after the military disasters of July 1863, when the Confederacy was headed down hill; it could hardly apply to the first two years of

---

[60] Ramsdell's review of Ulrich B. Phillips, *Life and Labor in the Old South* (Boston, 1929), in *Mississippi Valley Historical Review,* XVII (June 1930), 160-63. That review is included in this book.

the conflict.[61] While Ramsdell "tried to keep from saying anything positively offensive," he feared his constant "state of irritation" prevented success.[62]

Among nonprofessional historians who sought Ramsdell's counsel was George Fort Milton, who was writing a biography of Stephen A. Douglas. After *Eve of Confiict* appeared, Ramsdell wrote another reviewer that if he disliked the published work, he should have seen the manuscript. The Texan "wrote long protests and criticisms on nearly every chapter" that he read.

> I warned him that he was allowing the partisan character of the Douglas correspondence to betray him into a one-sided story and that the critics would certainly pitch into him about it. He promised to tone down his statements in certain particulars, and he has, but not enough.

Milton's "inside story of party politics" was an excellent contribution, and Ramsdell was

> temperamentally sympathetic with his central theme—that Douglas' plan offered

---

[61] Ramsdell's review of Frank L. Owsley, *State Rights in the Confederacy* (Chicago, 1925), in *Mississippi Valley Historical Review,* XIV (June 1927), 107-10. That review is included in this book.

[62] Ramsdell to Milo M. Quaife, January 2, 1926, in Ramsdell Papers. The reviewer added in his covering letter; "To have called attention to all the errors would have required a book as long as Owsley's."

> the only peaceful solution for the territorial squabble, weak as it was in logic, if only emotional strain joined to the inevitable maneuverings of party politics had permitted it to be honestly tried.[63]

Ramsdell's stock phrase appeared again: Milton "tried to prove too much." He "underestimated the strength of popular sentiment North and South"; overestimated Douglas' "southern popular support"; and understood neither the "local factional jealousies" that prompted Southern votes nor the Ultras who were committed to John C. Breckinridge.[64]

An appraisal of Ramsdell would be incomplete without some indication of his competence in the unrewarding task of refereeing papers submitted to scholarly reviews. Serving in that behind-the-scenes capacity for the *Journal of Southern History* in its early years, he demonstrated anticipated talent for separating papers with solid worth from others of questionable quality, and he readily recognized articles appropriate to newspaper Sunday supplements. Of one in the last

---

[63] Ramsdell to Dwight L. Dumond, February 2, 1935. See also Ramsdell to George Fort Milton, May 30, 1930, in Ramsdell Papers, in which he explained the limits of slavery expansion and the problem of slave labor in Virginia's mining and milling industries. Slaveowners could not afford to support slave families and employ only the men in heavy industrial work. Only when they combined agriculture and industry could slave labor be used profitably "in any large scale industry."

[64] Ramsdell's review of George Fort MIlton, *The Eve of Conflict: Stephen A. Douglas and the Needless War* (Boston, 1934), in *Mississippi Valley Historical Review,* XXII (June 1935), 105-106.

category, the critic observed that "the only suggestion of a new approach" was contained "in the rather cleverly phrased title."[65] But he was genuinely interested in the paper that showed promise even though it was not a finished product in research and craftsmanship. His constructive, critical evaluations—often running to a thousand words or more—enabled young scholars to revise·their work with improving effect.[66] His critical comments were often masterpieces of appraisal, and the managing editor might ponder whether historical scholarship would be better advanced by publishing the critiques rather than the articles.

If Ramsdell's critical thinking had a quality that distinguished it, *understanding* might be singled out as a special gift. He was not free of error—he might be wrong because he had covered all the evidence, even wrong in his speculative conclusions from the evidence before him. Yet it would be difficult to find in his thinking or writing either pettiness or sensationalism. He wore the historical ermine as befitted a man who conscientiously sought truth in appraising men who wrote history, or in presenting events and movements and men who formed its warp and woof. Perhaps one was surprised when he reported unfavorably on Owsley's *State Rights in the Confederacy* or viewed sympathetically Douglas S. Freeman's *R. E. Lee.*[67] In the first instance, he saw the

---

[65] Ramsdell to the writer, December 19, 1938, in writer's personal files.

[66] As examples, Ramsdell to the writer, April 17, 1935; April 25, 1937, *ibid* .

[67] Ramsdell's review of Douglas S. Freeman, *R. E. Lee: A Biography,* (4 vols., New York, 1934-1935), in *Journal of Southern History,* I (May 1935), 230-36. That review is

novice at work, misusing evidence to prove a thesis that had only limited validity; in the second, he observed the experienced master architect who built a magnificent house upon a firm foundation. He took no pride in the first, and regretted the assignment; he regarded the second as an opportunity for self-enlightenment.

The magnum opus on the Southern Confederacy was never completed; in fact, no chapters ever reached first draft. Readers of Ramsdell's briefer treatises, however, could visualize the nature of the Confederate history he would have written if time had permitted more years for the task. His paper on "Some Problems Involved in Writing the History of the Confederacy" presented at the first annual meeting of the Southern Historical Association in 1935, was a prospectus of the unfinished work. The problems were those that confronted Ramsdell. His discussion of them reveals his own comprehensive grasp of the Confederacy's main currents, his abundant knowledge of the documentary materials, his awareness of contributions scholars had made, and the innumerable gaps that awaited investigation before a definitive work could be written.

It was Ramsdell's habit to ask himself questions before he began to write, to weigh possible approaches, to state the broad plan, and to inventory his own equipment.

> . . . what we do *not* want [he began] is a history that lays undue emphasis upon any particular phase of the story, whether it be military operations, or the political and administrative policies and difficulties of

---

included in this book.

> the Confederate government, or the socio-economic conditions of the people.

On the contrary, what we do want is

> a full, comprehensive, well-balanced and articulated account that will give due weight to all discoverable factors in the struggle of the Southern people for independence and their failure to achieve it. Without sacrificing accuracy, it should have as much literary charm as the writer is capable of imparting to it.

The Confederacy's historian could not begin his narrative with secession, for almost every problem had antebellum roots. Thanks to Phillips, the plantation and slavery background was well known, but the role of the small farmers, the bankers, the merchants, the factors, and the industrialists had received scant attention. Unless the historian understood Southern businessmen and the South's "material resources," he could not hope to comprehend the problems that faced the Confederacy. There were few "searching studies of antebellum state politics." The historian would be severely handicapped if he did not understand "personal and factional rivalries," for "local antebellum political alignments and jealousies were carried over into the public affairs of the Confederacy."

In treating the war period, the historian could hardly ignore the Confederacy in battle array. Despite a tendency in the 1930's to relegate all military history to a subordinate position, "the fate of the 'revolution'" depended upon the success of the armies. The story of the supply services—the quartermaster's, subsistence, ordnance, and medical bureaus—was largely unknown. Wartime "mechanical industries," railroad transportation, financial problems, governmental measures that affected military operations, "the condition and attitudes of the people," were factors that must be weighed, and the historian would "find difficulty enough to challenge all his powers of analysis." Among the problems of government, the effective use of the Negro population and the disposition of such staples as tobacco and cotton would prove "troublesome." The influence of speculators and the blockade upon the Confederacy's fortunes, the privations endured by poor people, factionalism in Confederate politics, relief organizations, churches and hospitals, medical substitutes, foreign policy—these and countless other matters needed further investigation.

Ramsdell closed his prospectus with the significant observation that anyone who would "write a comprehensive history of the very complex life of the Confederacy must do a great deal of pioneer work for himself."[68] He was presenting problems of a quarter century ago. The monographic studies that have appeared in the interim have solved many of them; some others still await scholarly treatment.

Much has been written on "The Changing Interpretation of the Civil War" since Ramsdell presented his presidential address on that subject before the Southern Historical Association in 1936. The most

---

[68] Charles W. Ramsdell, "Some Problems Involved in Writing the History of the Confederacy," *ibid.*, II (May 1936), 133-47.

comprehensive of the studies, Thomas J. Pressly's *Americans Interpret Their Civil War,* surveyed with insight and understanding the conflicting viewpoints from the war period to the middle of the twentieth century. In briefer compass, Ramsdell viewed the evolution from apologias of contemporaries to the more objective accounts of the twentieth century's critical school. In presenting his conclusions, he differentiated between the two crises, secession and war. The first crisis, he thought, was inevitable, but the second was not—"unless those whose official positions gave them the power to choose war or peace were obliged to choose war for reasons of policy."[69] Admitting that both Southern whites and Negroes were "better off with slavery gone," he was convinced that the institution "would have begun to decline" soon after 1860 "and eventually have been abolished by the Southerners themselves." Despite "substantial progress" of Negroes since Appomattox, he questioned whether "their condition is better now than it would have been had a more orderly process of social and economic change been followed." He was also certain that the outcome of the war had not benefitted the "poor white" outcasts of the Old South. "This submerged class remains submerged."

Evaluating the results of the war, he found them wanting.

> Making all necessary allowances for our inability to weigh accurately the

---

[69] Ramsdell does not clarify this statement in the pages that follow, but at this point he cites his "Lincoln and Fort Sumter" paper, subsequently published *ibid.,* III (August 1937), 259-88.

> imponderables in the history of a great people [he wrote], can we say with conviction that this war accomplished anything of lasting good that could not and would not have been won by peaceful processes of social evolution? Is there not ground for the tragic conclusion that it accomplished little which was not otherwise attainable? Had the more than half a million lives and the ten billions of wasted property been saved, the wealth of the United States and the welfare of the people would not likely have been less than they are now. Perhaps some of the social and economic ills that have bedeviled us for the past fifty years would have been less troublesome. [70]

Ramsdell's speculations as to voluntary Southern termination of slavery and the easier adjustment of social and economic problems through evolution rather than revolution were accepted in some quarters, questioned in others. Some eyebrows were lifted higher at the thesis of his "Lincoln and Fort Sumter" study, which concluded that Lincoln, through ingenious strategy, "maneuvered the Confederates into firing the first shot in order that they, rather than he, should take the blame of beginning bloodshed." He resorted to the maneuver after deciding "that there was no other way than war for the salvation of

[70] 35Charles W. Ramsdell, "The Changing Interpretation of the Civil War," *ibid.*, III (February 1937), 1-25.

his administration, his party, and the Union."[71] If Ramsdell failed to prove his point convincingly, he exhibited rare talent for marshaling evidence and of piecing together the fragmentary contemporary information. In this premise the article has enduring merit as an example of meticulous inquiry, of search for obscure sources.[72]

In the absence of a comprehensive work on the war years, one must turn to Ramsdell's Fleming Lectures, *Behind the Lines in the Southern Confederacy,* for a basic interpretation of the Confederate States of America.

---

[71] Ramsdell, "Lincoln and Fort Sumter," *ibid.,* III (August 1937), 285. Near the beginning of the article he wrote: "Whether the war was inevitable, in any case, is a question that need not be raised here. It has been the subject of endless disputation and is one to which no conclusive answer can be given. But even though it be conceded that if the conflict had not arisen from the Fort Sumter crisis it would have sprung from some other incident growing out of the secession of the 'cotton states,' the actual firing of the 'first shot' placed the Southerners under a great moral and material disadvantage." *Ibid.,* 259.

[72] Unusual interest in the "Lincoln and Fort Sumter" article is indicated by excessive requests for reprints directed to the editorial office as well as to the author. "I expect a furious kick-back on this paper from various sources," Ramsdell wrote, "but I don't care as long as I believe I am right. I wish the thing were better written, however. Now that I have read it in type, I can see plenty of faults in the composition and in the presentation of the argument. But that has always been true of anything I have ever written." Of one modification in galley proof he said: "How in the world could I have left in the text such an asinine expression as 'strategic astuteness' I don't understand." Ramsdell to the writer, June 10, 1937, in writer's personal files. Many requests came also for Ramsdell's "The Changing Interpretation of the Civil War." Ramsdell to the writer, April 2, 1937, *ibid.*

Previous writers had indicated a variety of reasons for the failure of the stroke for independence—disparity in manpower, shortage of mechanical and military equipment, insufficient provisions, the Northern blockade, faulty strategy, monetary problems, internal dissension on the issue of state rights, inferior statesmanship, a psychological spirit of defeatism, lack of will to win.

If I were asked what was the greatest single weakness of the Confederacy [Ramsdell wrote], I should say, without much hesitation, that it was in this matter of finances. The resort to irredeemable paper money and to excessive issues of such currency was fatal, and it weakened not only the purchasing power of the government but also destroyed economic security among the people. In fact, there seems to be nothing vital that escaped its baneful influence.

On the other hand, Ramsdell recognized the causative arc as well as the apexed culmination. While hopelessness of Confederate finances was the main root of the difficulty, there were other contributing factors to collapse, among them "industrial weakness," illegal trade with the enemy, and the transportation problem, which he ranked "next to finances in its deleterious consequences."[73]

Ramsdell's last "great adventure," as he called it, was an edition of the *Laws and Joint Resolutions of the Last Session of the Confederate Congress (November 7, 1864-March 18, 1865), Together with the Secret Acts of Previous Congresses.* The volume, published the year

---

[73] Ramsdell, *Behind the Lines in the Southern Confederacy,* Ch. III.

before he died, exhibits Ramsdell as the careful, thorough, and precise editor, in search of the last scrap of evidence. The project was initiated by Duke University, which had acquired eighty-nine "manuscript enrolled acts and resolutions" of the Congress' final session, along with a few other acts and resolutions and "the official manuscript 'Register of Acts, C. S. A.'" The editor searched elsewhere for other missing acts and resolutions—Richmond newspapers, John H. Reagan Papers, and the National Archives—with rewarding discoveries.[74] Here was a volume of solid worth, assembled and edited in authoritative manner.

The historian's appraisals and articles of mature years manifested careful reading, cogent thinking, basic understanding, and clear expression. One of his pieces transcended usual competence. His review article, "Carl Sandburg's Lincoln," contributed to the *Southern Review,* indicates what Ramsdell could do with a theme that demanded perspective and inspired enthusiasm. The six-thousand-word essay was his last significant piece of writing. Craftsmanship attained literary distinction. Understanding promoted evaluation. Perspective assumed broad canvas. If only this essay survived, Ramsdell's mature historianship could be fairly evaluated. Sandburg and his Lincoln were mirrored against a mosaic of problems that faced the Civil War President as well as those that confronted the biographer. Of all the estimates of the poet's portrait, this one reveals better than any other the stature of the author, the subject—and the reviewer.

---

[74] Ramsdell (ed.), *Laws and Joint Resolutions of the Last Session of the Confederate Congress,* ix-xv.

In well-turned sentences, Ramsdell painted the mythical and legendary Lincoln whose vitality found expression in "thousands of books, pamphlets, essays, poems, and stories," in "references to him as an authority on countless disturbing questions, in appeals to his name in behalf of all sorts of causes." While

> one does not usually expect a critical biography from the hand of a poet who writes of his life-long hero, or an adequate understanding of all the great problems and developments which impinged upon his subject's career, the reader of these volumes soon discovers that Mr. Sandburg has tried to do an extremely difficult job with entire honesty. His warm imagination, his gift for vivid and colorful description—so welcome and so rare among the "professionals"—have not been allowed to run away with the evidence.

Sandburg had not escaped entirely the legend's "powerfully persuasive influence," for the poet "grew up in the midst of it" and "breathed its atmosphere all his life."

In words that command attention, the reviewer described Sandburg at his best.

> Everywhere are men and women talking of some aspect of the war and commending or condemning the doings of "Old Abe." All this is unlike the ordinary biography: it is

> more like a vast moving picture with thousands of actors passing in and out, back and forth across half a continent for a stage, but with one central figure who gradually draws the confusion into a certain order, gives it meaning and direction and finally dominates every scene. Little by little, almost imperceptibly, the personality of Lincoln changes too. Losing nothing of his native shrewdness or his deft political touch, he seems to rid himself of the awkward manners which dismayed so many early White House visitors, to lay aside more and more the crude jocosity which shocked others, and to reveal a deeper and profounder spiritual quality as he approached the ultimate tragedy awaiting him.

In one respect, at least, Ramsdell's estimate of Lincoln was more tenable than Sandburg's. For the poet, like so many other admirers of the emancipator's humanitarian qualities, disliked to paint him as a politician. While Sandburg did not ignore Lincoln's political genius, he seemed "less interested in the politician than in the statesman and less in the statesman than in the personality of the man." But, says the reviewer,

> Lincoln himself would never have denied that he was a politician; on the contrary, he would probably have been surprised and

> pained to learn that the distinction was withheld from him, for he delighted in his skill as a political manager and felt no squeamishness about the business of political trading for support. The truth is that he loved the game and made no pretense that he did not.

Ramsdell was among the first "professionals" to welcome Sandburg into the historical guild. Contrast if you will his appreciation for the poet's talent for revealing the real Lincoln with the cold reception accorded the *Prairie Years* of the mid-twenties. A reviewer characterized the two-volume biography of the pre-Civil War Lincoln as "a literary grab bag into which one may reach and draw out almost anything." Much of it was "imaginative poetry and historical fiction" punctuated by passages of "driveling sentimentality." Emphasizing factual errors, he concluded that "whatever else it may be, it is not history."[75] Ramsdell devoted some space to imperfections that marred many pages of *The War Years,* but he lifted understanding above the stuff of which history is made to the meaningful and accurate portrait the artist painted. Despite the imperfections, Sandburg produced a monumental work. The great interest in Lincoln was chiefly explained by "the peculiar qualities of the man himself," and Sandburg's biography presented "an unforgettable picture."[76] The professionals might

---

[75] Milo M. Quaife's review of Carl Sandburg, *Abraham Lincoln: The Prairie Years* (2 vols., New York, 1926) in *Mississippi Valley Historical Review,* XIII (September 1926), 287-91.
[76] Charles W. Ramsdell, "Carl Sandburg's Lincoln," *Southern Review* (Baton Rouge), VI (Winter 1941), 439-53.

profit from the lesson.

Historians may have had the "Sandburg's Lincoln" article at attention's focus when they lamented the loss of so much unwritten history with Ramsdell's passing. He seemed always on the verge of the great work that was never born. One explanation was his time-consuming university duties and many oddments that complicated his life. Unsparing attention to graduate students obstructed the even flow of scholarly activity. He wrote of "bedevilment" with doctoral orals, dissertations, and theses. Once he "quake[d] with fear and dread at the thought that a pile of them" might descend upon him at any moment; and again he planned to hibernate in New England where he would be inaccessible to students. His letters were punctuated with mild complaints of incidents and accidents, of "odds and ends that seem to hit me every day." An overzealous maid destroyed some note cards, and he himself mislaid some notes that occasioned searching again for references. An "emergency matter" robbed him of "time and sleep"; "energy that was far below par" delayed his writing. Even after a task was finished, he was interminably busy "tinkering with sentences and paragraphs."[77]

Lured by modest fees, he wrote introductions and encyclopedia articles,[78] wasted endless hours editing a

---

[77] For these complaints, see Ramsdell to the writer, May 29, 1935; April 1, July 2, December 14, 1936; March 1, June 13, 20, 30, 1937; December 9, 1938, in writer's personal files.
[78] As an example, Ramsdell wrote an introduction for the second and third volumes of the *Standard History of the World* (10 vols., New York, 1914), which·Paul Leland Haworth was revising a decade later. Ramsdell wrote to the editor: "Of course, I know nothing about the history of the period . . . but

history of Bell County, collaborated on a school text, and embarked on another—a joint project with J. G. de Roulhac Hamilton and Franklin L. Riley—that was never finished. Before the contract was canceled after a biennium, Ramsdell used some picturesque language. He wrote to Hamilton:

> If I ever get free of all the fool obligations I have gone into for the probability of making a little money, I promise myself that I shall spend the rest of my life in trying to live as I want to live—and that doesn't mean doing nothing.

A few weeks later he proposed to the same correspondent "a solemn compact and covenant . . . never to engage to write any future text book or any part thereof. This thing has taken all joy and freshness out of my young life." It would certainly "be the last 'pot boiler'" he would "ever undertake. I would much prefer to go back to the farm and raise pigs."[79]

---

as I would undertake to write a dissertation on Chinese history for $100.00, I have done what I could." He hoped that the two volumes would "never fall into the hands of any of my scholarly friends save yourself. Of you I expect discretion." Ramsdell to Haworth, June 8, 1926, in Ramsdell Papers.

[79] Ramsdell to J. G. de Roulhac Hamilton, January 26, February 20, April 13, 1918, in J. G. de Roulhac Hamilton Papers (University of North Carolina Library, Chapel Hill). See also, for this project, Ramsdell to Hamilton, January 9, 1917; Ramsdell to Franklin L. Riley, April 17, 1918, forwarded to Hamilton, *ibid.* The sad experience of 1917-1918 did not cure Ramsdell of the malady, as indicated in note immediately above.

These were some of the complexities and weaknesses that impeded Ramsdell's writing, but they were not unlike the autobiography of other scholars who published more. Perhaps he became more of a slave to routine and let it interfere unnecessarily with obligations to himself and his craft. Even if he had systematized his way of work, it is unlikely that he would have turned out book after book. He needed the pressure of program committees to prepare manuscripts, the urgent solicitation of prodding editors to submit them for publication. He lacked an impelling obsession to speed his pen.

Two assumptions seem valid: first, that his storehouse held much historical knowledge in reserve; and, second, that the articles and reviews and books he put in print merely illustrated his comprehension of the critical years at mid-nineteenth century. The unwritten part of his learning, as well as his wise counsel and official services, contributed to his "deanship" of Southern historians. How shall we evaluate his stature in the historical world?

First, he and his colleagues assembled a notable collection of printed, manuscript, and microfilm records at the University of Texas, making it a center where scholars may investigate nineteenth-century Southern history.

Second, he gave liberally of his time to students who profited from his understanding of history and the congenial atmosphere of classroom, seminar, and conference. The investment yielded returns in competently trained students, some of whom became reputable historians. His influence as a teacher transcended formal instruction to include a widely scattered clientele whose relationships were informal and casual.

Third, he developed a personal philosophy of history based upon understanding as a major ingredient. While there was nothing new in the concept, it enabled him in teaching and writing to view history in terms of human complexity that no central theme or neat pattern could explain. Human behavior, whether of the individual or of collective society, was intricate and chaotic, and no simple explanation could account for a people's development.

Fourth, he expanded the frontiers of history by distilling a segment of it, more significant in quality than in quantity.

Finally, he bequeathed a body of challenging interpretations which are worthy of consideration by students who seek meaning in past events. Not all of them were original, but his analyses of basic historical survivals were a worthy contribution. Among them were the following:

> The institution of slavery had reached its natural limits before the crisis of 1860 arrived. Ultimate extinction was its destiny.

The Civil War was not inevitable. It could have been avoided if an artificial issue—expansion of slavery into territory where natural causes excluded it—had not reached the stage of acrimonious political debate.

The hopeless condition of Confederate finances was a major factor that led to Appomattox. It occupied a central position on the causative arc.

Abraham Lincoln was an adroit politician whose growth under pressure of responsibility was a deciding factor in the struggle to preserve the Union. Evidence of this quality was apparent as early as the firing on Fort Sumter.

The results of the war were negative. The submerged elements in the Southern population—black and white—would be further advanced in the twentieth century if improved social and economic status had arrived through the processes of evolution. A half million lives and billions of dollars in property were needlessly sacrificed.

This summarization suggests the validity of Barker's characterization of Ramsdell in 1913 as a "conservative progressive." The atmosphere of his formative years in Reconstruction's aftermath was conservative. The Texan's concept of history was stability tempered by change, or, as Ramsdell phrased it, "continuous evolution." He clung tenaciously to the fundamentals of the old while incorporating the new. The historian of Reconstruction in Texas who sought objectivity without attaining it advanced far along the road to liberal understanding in appraising Carl Sandburg and his Abraham Lincoln. Historiography to Ramsdell was sometimes confirmatory, often transforming.

Part Two

Ramsdell's Famous Treatises

Lincoln and Fort Sumter[80]

by

Charles W. Ramsdell

We are to have civil war, if at all,
because Abraham Lincoln loves a party
better than he loves his country. . . .
Mr. Lincoln saw an opportunity to
inaugurate civil war without appearing
in the character of an aggressor.

"WHY?"
Providence (R.I.) Daily Post
The day after the commencement of
the bombardment of Fort Sumter
April 13, 1861

When the Confederate batteries around Charleston Harbor opened fire on Fort Sumter in the early morning hours of April 12, 1861, they signaled the beginning of the most calamitous tragedy in the history of the American people. Because the Confederate authorities ordered the attack it is generally held that they were directly responsible for the horrors of the ensuing four years. Certainly that was the feeling in the North, then and afterwards, and it became the verdict of austere historians.

Whether the war was inevitable, in any case, is a

---

[80] Charles W. Ramsdell, "Lincoln and Fort Sumter," *The Journal of Southern History,* Vol. 3, Issue 3 (August, 1937), 259 - 288. Article and citation are verbatim.

question that need not be raised here. It has been the subject of endless disputation and is one to which no conclusive answer can be given. But even though it be conceded that if the conflict had not arisen from the Fort Sumter crisis it would have sprung from some other incident growing out of the secession of the "cotton states," the actual firing of the "first shot" placed the Southerners under a great moral and material disadvantage. The general Northern conviction that the "rebels" had made an unprovoked attack upon the little Federal garrison added thousands of volunteers to the Union armies and strengthened the determination of the Northern people to carry the struggle through to the complete subjugation of the South.

The Confederate leaders who ordered the bombardment were not vicious, feeble-minded, irresponsible, or inexperienced men. As even a casual investigation will show, they had been fully aware of the danger of taking the initiative in hostilities and had hoped for peace. How then could they be so blind as to place themselves at this manifest disadvantage?

The story of the development of the Fort Sumter crisis has been told many times, but it is so full of complexities that there is little wonder that many of its most significant features have been obscured with a resultant loss of perspective. On the one hand, most accounts have begun with certain assumptions which have affected the interpretation of the whole mass of evidence; on the other, too little credit has been given to Abraham Lincoln's genius for political strategy, which is truly surprising in view of all the claims that have been made for the abilities of that very remarkable man. The purpose of this paper is to place the facts already known in their logical and chronological order and to re-evaluate them in that setting in the belief that when thus arranged they will throw new light upon this momentous affair.

The early stages of the Sumter problem can be dealt with in summary form. It is well known that six days after the secession of South Carolina Major Robert Anderson, who had been stationed at Fort Moultrie in command of all the United States forces in Charleston Harbor, abandoned Moultrie and moved his command into the new and still unfinished Fort Sumter where he thought his force would be better able to resist attack. The South Carolina authorities evidently had had no intention of attacking him for they thought they had an understanding with President Buchanan for maintaining the military status quo; but they immediately occupied Fort Moultrie and Castle Pinckney and made protest to Buchanan, demanding that Anderson be sent back to Moultrie. Buchanan refused to admit their ground of protest or to order Anderson back; then early in January he ordered relief to be sent secretly to the garrison on a merchant steamer. This vessel, *The Star of the West,* was forced back from the entrance of the harbor by the military authorities of the state, and the South Carolinians were with some difficulty restrained by the leaders in other Southern states from assaulting Fort Sumter. Thereafter Buchanan refrained from the use of force, partly because Anderson insisted that he was in no danger, partly because he still hoped for some peaceful adjustment, if not by Congress itself, then by the Peace Conference which was soon to assemble in Washington, and partly because he was averse during the last weeks of his term to beginning hostilities for which he was unprepared.

By February 1 six other cotton states had passed ordinances of secession and each of them, as a matter of precaution and largely because of the happenings at Charleston, seized the forts, arsenals, customs houses, and navy yards within its own borders. There were two exceptions, both in Florida. Fort Taylor, at Key West, was left undisturbed; and Fort Pickens, at the entrance of Pensacola Bay and on the extreme western tip of Santa Rosa Island, was occupied by a small Federal force much as Fort Sumter had been.

Since Fort Pickens plays a part in the development of the Sumter crisis, some explanation of the situation at that point becomes necessary. In the beginning this fort was not occupied by troops, but a company of artillery, under Lieutenant Adam J. Slemmer, was stationed at Barrancas Barracks, across the neck of the bay about a mile and a half to the north of Pickens, and close by the Navy Yard. The town of Pensacola was some six miles farther up the bay. On January 10 Lieutenant Slemmer, hearing that the governors of Florida and Alabama were about to send troops to seize the forts and the Navy Yard and in accordance with instructions from General Winfield Scott, removed his small command to Fort Pickens. On the twelfth the Navy Yard capitulated to the combined state forces under Colonel W. H. Chase. Chase then demanded the surrender of Fort Pickens, which Slemmer refused. After some further correspondence between the two opposing officers, a group of nine Southern senators in Washington, on January 18, urged that no attack should be made on Fort Pickens because it was "not worth a drop of blood."[81] These senators believed that the Republicans in Congress were hoping to involve 'the Buchanan administration in hostilities in order that war might open before Lincoln's inauguration. On January 29 an agreement was effected at Washington by Senator Stephen R. Mallory of Florida, and others, with President Buchanan and his secretaries of War and the Navy to the effect that no reinforcement would be sent to Fort Pickens and no attack would be made upon it by the secessionists.[82] The situation at Fort Pickens then became somewhat like that at Fort Sumter; but there were certain differences. Fort Pickens did not threaten the town of Pensacola as Fort Sumter did Charleston; it was easily accessible from the sea if reinforcements should be decided upon; and there was no such excitement over its continued occupation by United States troops as there was about Sumter. As soon as the new Confederate government was organized the Confederate Congress, on February 12, by resolution took charge of "questions existing between the several States of this Confederacy and the United States respecting the occupation of forts, arsenals, navy yards and other public establishments." This hurried action was taken in order to get the management of the Sumter question out of the hands of the impatient and rather headlong Governor Francis W. Pickens of South Carolina, who, it was feared, might precipitate war at any time.[83] In fact, the public mind,

---

[81] *The War of the Rebellion : A Compilation of the Official Records of the Union and Confederate Armies* (Washington, 1880-1901), Ser. I, Vol. I, 445-46. Hereafter cited as *Official Records.*

[82] *Ibid.*, 355-56.

[83] *Journal of the Congress of the Confederate States of*

North and South, sensed accurately that the greatest danger to peace lay in Charleston Harbor.

This danger, of course, was in the irreconcilable views of the two governments concerning their respective claims to the fort. To the Washington officials Sumter was not merely the legal property of the Federal government; its possession was a symbol of the continuity and integrity of that government. To withdraw the garrison at the demand of the secessionists would be equivalent to acknowledging the legality of secession and the dissolution of the Union. There was also, especially with the military officials, a point of honor involved; they could not yield to threats of force. The attitude of the Southerners was based upon equally imperative considerations. In their view the Confederate States no longer had any connection with the government on the Potomac; they were as independent as that other seceded nation, Belgium. No independent government could maintain its own self-respect or the respect of foreign governments if it permitted another to hold an armed fortress within the harbor of one of its principal cities. When South Carolina had ceded the site for the fortification it had done so for its own protection. That protection was now converted into a threat, for the guns of Sumter dominated not only every point in the harbor but the city of Charleston itself. We may conceive an analogous situation by supposing that Great Britain at the close of the American Revolution had insisted upon retaining a fortress within the harbor of Boston or of New

---

America, 7 vols. (Washington, 1904-1905), I, 47; Samuel W. Crawford, *The Genesis of the Civil War: The Story of Sumter* (New York, 1887), 261-62.

York. The Confederate government could not, without yielding the principle of independence, abate its claims to the fort.

During the last six weeks of Buchanan's term the situation at Charleston remained relatively quiet. Anderson and his engineers did what they could to strengthen the defenses of Sumter; while the state and Confederate officers established batteries around the harbor both to repel any future relief expedition and, in case of open hostilities, to reduce the fort. Although Governor Pickens had wished to press demands for surrender and to attack the fort if refused, he had first sought the advice of such men as Governor Joseph E. Brown of Georgia and Jefferson Davis of Mississippi. Both advised against any such action, partly because they still had some hope of peace and partly because they saw the danger of taking the initiative.[84] Although Anderson was under constant surveillance, he was allowed free use of the mails and was permitted to purchase for his men fresh meats and vegetables in the Charleston market. Other necessities, which under army regulations he must procure from the regular supply departments of the army, he was not allowed to receive because that would be permitting the Federal government to send relief to the garrison and involve an admission of its right to retain the fort. Anderson consistently informed the authorities at Washington during this time that he was safe and that he could hold out indefinitely. The Confederate government, having taken over from the state all negotiations

---

[84] Crawford, *Genesis of the Civil War,* 263-68; Dunbar Rowland (ed.), *Jefferson Davis, Constitutionalist,* 10 vols. (Jackson, Miss., 1923), V, 36-37, 39-40.

concerning the fort, was moving cautiously with the evident hope of avoiding hostilities. On February 15 the Confederate Congress by resolution requested President Davis to appoint three commissioners to negotiate with the United States "all questions of disagreement between the two governments" and Davis appointed them on February 25.[85] They reached Washington on March 5, the day after Lincoln's inauguration.

Southern as well as Northern men waited anxiously to learn what policy would be indicated by the new President of the United States in his inaugural address. It is not necessary to dwell long on what Abraham Lincoln said in that famous paper. He stated plainly that he regarded the Union as unbroken, the secession of the seven cotton states as a nullity. In this he merely took the position that Buchanan had taken. He also said that he would enforce the laws of the Union in all the states; but he immediately softened this declaration by saying that he would not use violence unless it should be forced upon the national authority. Then he added, "The power confided to me will be used to hold, occupy and possess the property and places belonging to the government, and to collect the duties and imposts; but beyond what may be necessary for these objects, there will be no invasion, no using of force against or among the people anywhere." And later on: "In your hands, my dissatisfied fellow countrymen, and not in mine, is the momentous issue of civil war. The government will not assail you. You can have no conflict without being yourselves the aggressors." How is it possible to reconcile the declaration that he

---

[85] *Journal of the Congress of the Confederate States,* I, 46, 52, 55, 85-86.

would occupy "the property and places belonging to the government" with the promise that the government would not assail his dissatisfied fellow countrymen who either held or claimed the right to those places? While ostensibly addressing the Southerners, was he really directing these last soothing words to the anxious antiwar elements in the North? Although it is improbable that he had this early any definite plan in mind, his warning that the secessionists would be the aggressors, if civil war should come, may be significant in view of what he was to be engaged in exactly a month from that day.

But the inaugural should not be regarded as the declaration of a definite program; for while the new President was careful to lay down the general principle that the Union was legally unbroken, he refrained with equal care from committing himself to any course of action. If he hedged at every point where a statement of active policy was expected, it was because he could not know what he would be able to do. Caution was necessary; it was not merely political expediency, it was at that juncture political wisdom. Cautious reticence, until he knew his way was clear, was a very marked trait of Abraham Lincoln.[86] There is another characteristic

---

[86] This characteristic of Lincoln was attested to by numbers of his associates, sometimes with evident irritation. W. H. Herndon once wrote, "He was the most secretive—reticent—shut-mouthed man that ever lived." Herndon to J. E. Remsburg of Oak Mills, Kansas, September 10, 1887 (privately printed by H. E. Baker, 1917). See also A. K. McClure, *Lincoln and Men of War-Times* (Philadelphia, 1892), 64-68, for statements of Leonard Swett, W. H. Lamon, A. K. McClure, and David Davis. Judge Davis said, "I knew the man well; he was the most reticent, secretive man I ever saw or expect to see."

quality in this address. Lincoln had developed an extraordinary skill in so phrasing his public utterances as to arouse in each special group he singled out for attention just the reaction he desired. To the extreme and aggressive Republicans the inaugural indicated a firm determination to enforce obedience upon the secessionists; to the Northern moderates and peace advocates, as well as to the anxious Unionists of the border slave states, not yet seceded, it promised a conciliatory attitude; in the seceded states it was interpreted as threatening coercion and had the effect of hastening preparations for defense. In the latter part of the address Lincoln had counseled the people generally to avoid precipitate action and to take time to think calmly about the situation. He doubtless hoped to be able to take time himself; but he discovered within a few hours that there was one problem whose solution could not be long postponed. On the very day of his inauguration Buchanan's secretary of war, Joseph Holt, received a letter from Major Anderson in which for the first time the commander at Fort Sumter expressed doubt of his ability to maintain himself. More than this, Anderson estimated that, in the face of the Confederate batteries erected about the harbor, it would require a powerful fleet and a force of twenty thousand men to give permanent relief to the garrison. Since it was his last day in office, Buchanan had the letter referred to Lincoln; and when on March 5 Holt submitted it to the new President he accompanied it with a report sharply reviewing Anderson's previous assurances of his safety.[87] Lincoln called General Scott

---

[87] Anderson's letter has not been located, but see *Official Records,* Ser. I, Vol. I, 197-202. For Holt's letter, Horatio King,

into conference and the General concurred with Anderson. After a few days of further consideration Scott was of the same opinion and was sustained by General Joseph G. Totten, chief of the Army Engineers. These men considered the question primarily as a military problem, although Scott could not refrain from injecting political considerations into his written statement. In doing this the aged General was suspected of following the lead of Secretary William H. Seward who was already urging the evacuation of Sumter, in order to avoid precipitating hostilities at that point, and the reinforcement of Fort Pickens in order to assert the authority of the government. Lincoln accepted at least a part of Seward's plan, for on March 12, General Scott, by the President's direction, sent an order to Captain Israel Vogdes, whose artillery company was on board the U. S. Steamer *Brooklyn,* lying off Fort Pickens, directing him to land his company, reinforce Pickens, and hold it. Instead of sending the order overland, Scott sent it around by sea with the result that it did not reach its destination until April 1, and then the navy captain in command of the ship on which the artillery company was quartered refused to land the troops because the orders from the former Secretary of the Navy directing him to respect the truce with the Confederates had never been countermanded. The fort was not reinforced at that time, a fact of which Lincoln remained ignorant until April 6. We shall return to the Fort Pickens situation later.

Meanwhile Lincoln was considering the Fort Sumter problem. He had learned that Anderson's supplies were running short and that the garrison could not hold out

---

Turning on the Light (Philadelphia, 1896), 126-28.

much longer without relief. Although both General Scott and General Totten had advised that the relief of the fort was impracticable with the forces available, Gustavus V. Fox, a former officer of the navy and a brother-in-law of Postmaster-General Montgomery Blair, believed that it would be possible to reach the fort by running small steamers past the Confederate batteries at the entrance to the harbor. Fox had first proposed this to Scott early in February; he now came forward again with the backing of Montgomery Blair and presented his plan and arguments to Lincoln on March 13. The President seems to have been impressed, for on March 15 he asked for the written opinions of his cabinet on the question whether, assuming that it was now possible to provision Sumter, it was wise to attempt it. All, save Montgomery Blair, advised against an expedition.[88] Apparently this overwhelming majority of his cabinet at first decided him against the plans, for there is considerable evidence, although it is not conclusive, that he was about to order Anderson to evacuate. Certainly rumors of impending orders for evacuation were coming from various high official circles in Washington, aside from those for which Seward seems to have been responsible.[89] There is the familiar story of how old Frank Blair, brought to the

---

[88] Secretary Chase favored a relief expedition, but only if it would not bring on an expensive war, a position that was so equivocal that he can hardly be said to stand with Montgomery Blair. John G. Nicolay and John Hay (eds.), *Abraham Lincoln: Complete Works,* 2 vols. (New York, 1894), II, 11-22, for replies of the cabinet.

[89] The newspapers carried these reports almost every day and the belief in their accuracy seems to have been general, even among the war faction of the Republicans.

White House by his son Montgomery, found the President about to sign the evacuation order and protested so vigorously that Lincoln did not sign it.

Lincoln now found himself facing a most difficult and dangerous situation and the more he considered it the more troublesome it appeared. It seems reasonably certain that he never wanted to give up Sumter. As early as December 24, 1860, having heard a wild rumor that the forts in South Carolina were to be surrendered by the order or consent of President Buchanan, he had written from Springfield to Senator Lyman Trumbull that he would, "if our friends at Washington concur, announce publicly at once that they are to be retaken after the inauguration."[90] Af ter he had arrived at Washington and had taken up the burden of office he saw that the problem was not so simple as it had looked from the frontier town of Springfield. His Secretary of State, a man of far greater political experience than himself, was urging him to make his stand for the authority of the government at Fort Pickens and not at Sumter, for Seward could not see how it would be possible to reinforce Sumter without putting

[90] Gilbert A. Tracy, *Uncollected Letters of Abraham Lincoln* (Boston and New York, 1917), 173. Lincoln had written "confidentially" to Major David Hunter on December 22, "If the forts fall, my judgment is that they are to be retaken." A. B. Lapsley (ed.), *The Writings of Abraham Lincoln,* 8 vols. (New York, 1905-1906), V, 199. It will be remembered that the original draft of the inaugural had contained a declaration that he would "reclaim the public property and places which have fallen," but that this was changed at the suggestion of Orville H. Browning to a more general and less threatening statement. John G. Nicolay and John Hay, *Abraham Lincoln, A History,* 10 vols. (New York, 1886-1892), III, 319, 333-34, n. 12.

the administration in the position of the aggressor. That would be a fatal mistake. Fort Pickens, on the other hand, could be relieved from the Gulf side without coming into direct conflict with the Confederates.

It would be extremely interesting to know what was passing through Lincoln's mind during those difficult days when, bedeviled by importunate office seekers, he could find little time for considering what he should do about the re-establishment of Federal authority in the seceded states and especially about the imperiled fort at Charleston. As was his habit, he left few clues to his reflections and it is impossible to say with assurance what ideas he picked up, examined, and discarded. One plan which he seems to have entertained for a short while, just after the adverse cabinet vote on relieving Sumter, contemplated the collection of customs duties on revenue vessels, supported by ships of war, just outside the Confederate ports; and there were hints in the press that Anderson's force was to be withdrawn to a ship off Charleston. If it were seriously considered, the plan was soon abandoned, possibly because of legal impediments or more probably because it did not fully meet the needs of the situation.[91] But although Lincoln kept his thoughts

---

[91] Lincoln to Chase, Welles, and Bates, March 18, 1861, in Nicolay and Hay (eds.), *Lincoln: Works*, II, 24-25. The Morrill tariff, passed in February, had raised rates far above the former ones while the Confederate Congress had enacted a low tariff. The difference in rates was causing anxiety to Northern importers and shippers, and also to the administration, lest it deflect imports to the South and stimulate smuggling across the new border to the great injury of the Northern ports and the loss of customs receipts. The tariff differential might even swing some of the border states over to the Confederacy. The New York *Times* was greatly

disturbed at the prospect and roundly condemned the Morrill tariff. The issues of the *Times* for March 13, 15-20, and 22, intimated that the President was considering the above-mentioned plan. The legal impediments seem to have consisted in the absence of any law of Congress permitting such a procedure and the nonexistence of local Federal courts for the adjudication of cases arising out of the enforcement of the revenue laws. This tariff question may have had more influence upon the final determination of Lincoln's policy that the evidence now available shows. **[COMPILER'S NOTE: Ramsdell is correct. The Morrill Tariff, which threatened to reroute Northern shipping almost overnight away from the high tariff North and into the low tariff South (compare the 47 to 60% Morrill Tariff of the North with the South's 10% tariff), together with the loss of the North's captive Southern manufacturing market, was the one-two punch that promised the total annihilation of the Northern economy and made war necessary from Lincoln's standpoint. No country can survive the loss of its manufacturing industry AND shipping industry, overnight, and not collapse. It was not just the loss of wealth that was foreshadowed. It was runs on banks, massive unemployment, people in the street because they can't feed their families or protect their women. That's what was coming fast to the North. Henry L. Benning, a justice on the Georgia Supreme Court before the war and later one of Robert E. Lee's most able brigadier generals (for whom Fort Benning, Georgia is named) stated before the war: "The North cut off from Southern cotton, rice, tobacco, and other Southern products would lose three fourths of her commerce, and a very large proportion of her manufactures. And thus those great fountains of finance would sink very low. . . . Would the North in such a condition as that declare war against the South?" The answer is YES, and that is proven by Lincoln's actions and the calls for war by business people in the North who were terrified and furious watching their wealth disappear**

to himself he must have studied public opinion closely, and we may be able to follow his thinking if we examine for ourselves the attitudes of the several groups in the North as they revealed themselves in those uncertain days of March.

It must not be forgotten that, notwithstanding Lincoln's smashing victory in the free states in November, his party was still new and relatively undisciplined. His support had come from a heterogeneous mass of voters and for a variety of reasons. The slavery issue, the drive for a protective tariff and internal improvements, the promise of free homesteads in the West, and disgust at the split among the Democrats had each played its part. Many voters had been persuaded that there was no real danger of a disruption of the Union in the event of his election. The secession of the border states had now thrown the former issues into the background and thrust to the front the question whether the discontented Southerners should be allowed to depart in peace or whether the government should, as Lincoln phrased it, "enforce the law" and in so doing bring on war with the newly formed Confederacy. As always, when a new and perilous situation arises, the crosscurrents of public opinion were confusing. As Lincoln, pressed on all sides, waited while he studied the drift, he could not fail to note that there was a strong peace party in the North which was urging the settlement of difficulties without resort to force. On the other hand the more aggressive party men

---

and their economy descend into anarchy. War would solve all of that including Lincoln's enormous political problems, which threatened to obliterate the Republican Party and destroy Lincoln's administration at the outset.]

among the Republicans, to whom he was under special obligations, were insisting that he exert the full authority of the government even to the extent of war. This group included some of the most active and powerful members of his party whom he could not afford to antagonize. One disturbing factor in the situation was the marked tendency of many voters who had supported him in November to turn against the Republicans, as was shown in a number of local elections in Ohio and New England. While the peace men attributed this reversal to fear of war, the more aggressive Republicans insisted that it was caused by disgust at the rumors that Fort Sumter would be given up to the secessionists.[92] Reinforcing the Northern conservatives were the majorities in the eight border slave states who had thus far refused to secede but who were openly opposed to any "coercive" action against their brethren in the Lower South. The Virginia State Convention, which had convened on February 13 and was in complete control of the conditional Unionists, was still

---

[92] These elections were not actually held until April 1 in Ohio and Connecticut and April 3 in Rhode Island, but the pre-election evidences of defection had greatly alarmed the Republicans in the latter part of March. The fusion of the Democrats and other "Union savers" carried all the larger cities of Ohio, defeated two radical Republican congressmen in Connecticut, re-elected Governor William Sprague in Rhode Island, and won a majority of the legislature in that state. Cincinnati *Commercial,* April 3, 1861; Columbus (Ohio) *Crisis,* April 4, 1861; New York *Times,* March 30, April 2, 4, 1861; J. H. Jordan to S. P. Chase, March 27, J. N. and J. B Antram to Chase, April 2, and W. D. Beckham to Chase, April 2, 1861, in Chase Papers, Library of Congress. I am indebted to Mrs. W. Mary Bryant of the University of Texas for copies of these letters.

in session, evidently awaiting his decision. Therefore, if he should adopt a strongly aggressive policy he might find himself opposed by the large group of peace men in the North while he precipitated most if not all of the border slave states into secession and union with the Confederacy.[93] If, on the other hand, he failed to act decisively, he was very likely to alienate the radical Republicans who were already manifesting impatience. In either case he would divide his party at the very beginning of his administration and increase the risk of utter failure. There was, however, some cheering evidence among the business elements of a growing irritation against the secessionists because of the depression which had set in with the withdrawal of South Carolina; and if the Confederates should add further offense to their low tariff policy or adopt more aggressive tactics with respect to the forts, this feeling might grow strong enough to overcome the peace men.

He had promised to maintain the Union, but how was he to attempt it without wrecking his chances at the very outset? It was now too late to restore the Union by compromise because, having himself rejected all overtures in December, he could not now afford to offer what he had recently refused. Moreover, there was no indication that the Confederates would accept at this late date any compromise he might proffer. He must do something, for the gradual exhaustion of the supplies of the garrison in Fort Sumter would soon force his hand. He could not order Anderson to evacuate without

[93] There are some indications, however, that at this time Lincoln overestimated the Unionist strength in the border slave states.

arousing the wrath of the militant Unionists in the North. If he continued to let matters drift, Anderson himself would have to evacuate when his supplies were gone. While that would relieve the administration of any charge of coercion, it would expose the government to the accusation of disgraceful weakness and improve the chances of the Confederacy for foreign recognition.[94] If he left Anderson to his fate and made ostentatious display of reinforcing Fort Pickens, as Seward was urging him to do, would he gain as much as he lost? Was it not best, even necessary, to make his stand at Sumter? But if he should try to relieve Anderson by force of arms, what was the chance of success? Anderson, supported by the high authority of General Scott, thought there was none. If, as Captain Fox believed, swift steamers could run the gauntlet of the Confederate batteries and reach the fort with men and supplies, would they then be able to hold it against attack? Failure in this military movement might seriously damage the already uncertain prestige of the administration. Would it not be looked upon as aggressive war by the border state men and perhaps by the peace men in the North? Could he risk the handicap

---

[94] Lincoln's special message to Congress, July 4, 1861, indicates that he had weighed some of these considerations. "It was believed, however, that to abandon that position [Sumter] under the circumstances would be utterly ruinous; that the *necessity* under which it was to be done would not be fully understood; that by many it would be construed as a part of a *voluntary* policy; that at home it would discourage the friends of the Union, embolden its adversaries, and go far to insure to the latter a recognition abroad; that, in fact, it would be our national destruction consummated. This could not be allowed." J. D. Richardson (comp.), *Messages and Papers of the Presidents*, 10 vols. (Washington, 1896-1899), VI, 21.

of appearing to force civil war upon the country? In every direction the way out of his dilemma seemed closed.

There was one remote possibility: the Confederates themselves might precipitate matters by attacking Sumter before Anderson should be compelled to evacuate by lack of supplies. But the Confederates, though watchful, were showing great caution. General P. G. T. Beauregard, in command at Charleston since March 6, was treating Major Anderson with elaborate courtesy. The government at Montgomery was in no hurry to force the issue, partly because it was quite well aware of the danger of assuming the aggressive and partly because it was waiting to see what its commissioners would be able to effect at Washington, where Seward was holding out hopes to them of the eventual evacuation of Sumter. At some time, while turning these things over in his mind, this daring thought must have occurred to Lincoln: Could the Southerners be *induced* to attack Sumter, to assume the aggressive and thus put themselves in the wrong in the eyes of the North and of the world?[95] If they could, the

---

[95] It would be most surprising to find that such an idea never occurred to Lincoln, since not only were many Republicans suggesting it as a possibility, but various Republican newspapers were constantly reiterating the suggestion that if any clash came the secessionists would be responsible. The predictions of the newspapers may have been "inspired," but if so, that fact makes it more certain that the idea was being discussed in the inner circles of the administration. J. H. Jordan wrote Chase form Cincinnati, March 27, "In the name of God! why not hold the Fort? Will reinforcing & holding it cause the rebels to attack it, and thus bring on 'civil war'? What of it? That is just what the government ought to wish to bring about, and ought to do all it can . . . to bring about. Let them attack the Fort, if they will—it will then be *them* that

latent irritation perceptible among the Northern moderates might flame out against the secessionists and in support of the government. The two wings of his party would unite, some at least of the Democrats would come to his support, even the border-state people might be held, if they could be convinced that the war was being forced by the secessionists. Unless he could unite them in defense of the authority of the government, the peaceable and the "stiff-backed" Republicans would split apart, the party would collapse, his administration would be a failure, and he would go down in history as a weak man who had allowed the Union to crumble in his hands. As things now stood, the only way by which the Union could be restored, his party and his administration saved, was by an unequivocal assertion of the authority of the government, that is, through war. But he must not openly assume the aggressive; that must be done by the secessionists. The best opportunity was at Fort Sumter, but the time left was short for Anderson was running short of essential supplies.

Let us examine closely what Lincoln did after the middle of March, taking care to place each movement as nearly as possible in its exact sequence. We have seen that Captain Fox made his argument to Lincoln for a combined naval and military expedition on March 13 and that the cabinet, with the exception of Montgomery Blair and the equivocal Chase, had voted against it on the fifteenth. Fox then offered to go in person to Fort Sumter to investigate the situation and Lincoln gave him

---

commenced the war." The general idea of such an outcome was in the air; the contribution of Lincoln himself was the maneuver by which this desirable solution was brought about.

permission. He arrived in Charleston on March 21 and was allowed to see Anderson that night. He and Anderson agreed that the garrison could not hold out longer than noon of April 15. Although Anderson seems to have remained unconvinced of its feasibility, Fox returned to Washington full of enthusiasm for his plan.

On the very day that Fox arrived in Charleston, Lincoln had dispatched to that city a close friend and loyal supporter, Ward H. Lamon, a native of Virginia and his former law partner in Illinois. This sending of Lamon on the heels of Fox is an interesting incident. The precise nature of his instructions has never been fully revealed. Lamon himself, in his *Recollections*, merely says he was sent "on a confidential mission" and intimates that he was to report on the extent of Unionist feeling in South Carolina. He arrived in Charleston on the night of Saturday, March 23; visited James L. Petigru, the famous Unionist, on Sunday and learned from him that there was no Unionist strength in the state, that "peaceable secession or war was inevitable"; and on Monday morning obtained an interview with Governor Pickens. In reply to questions the Governor stated very positively that any attempt on the part of President Lincoln to reinforce Sumter would bring on war, that only his "unalterable resolve *not* to attempt any reinforcement" could prevent war. Lamon, whether through innocence or guile, left the impression with the Governor, and also with Anderson whom he was permitted to visit, that the garrison would soon be withdrawn and that his trip was merely to prepare the way for that event. He left Charleston on the night of the twenty-fifth, arrived in Washington on the twenty-seventh, and reported to Lincoln what he had learned.[96] What had he been sent to Charleston to do? There must have been some purpose and it could hardly have been to prepare the way for Anderson's evacuation.[97] Does it strain the evidence to suggest that it was chiefly to find out at first hand how strong was the Southern feeling about relief for Fort Sumter and that this purpose was camouflaged by the vague intimations of evacuation? But it is quite probable that Lamon himself did not understand the real purpose, for it is altogether unlikely that the cautious Lincoln would have divulged so important a secret to his bibulous and impulsive young friend. But if there was such an ulterior purpose, Lincoln now had the information directly from Charleston that any sort of relief would result in an attack upon the fort.

According to Gideon Welles, whose account of these events was written several years later, Lincoln sometime in the latter half of March had informed the members of his cabinet that he would send relief to Sumter. During a cabinet meeting on March 29 (two days after Lamon's return), when the matter was again discussed, Lincoln, at

---

[96] Ward H. Lamon, *Recollections of Abraham Lincoln* (Washington, 1911), 68-79.

[97] On April 1 Lincoln sent word, through Seward, to Justice John A. Campbell that Lamon had no authority to make such a promise. Not only that but, according to the same source, he stated that "Lamon did not go to Charleston under any commission or authority from Mr. Lincoln." Henry G. Connor, *John Archibald Campbell* (Boston and New York, 1920), 127. The words "commission or authority" may have been a mere technical evasion of responsibility, for Lamon himself recounts the conversation between Lincoln, Seward, and himself when Lincoln asked him to go. It is possible, of course, that Justice Campbell misunderstood the exact language or meaning of Seward.

the suggestion of Attorney General Edward Bates, again requested each member to give his opinion in writing on the question of relieving Sumter. Whether Lincoln's known determination, political pressure, or some other influence had effected it, there was a marked change from the advice given just two weeks earlier. Now only Seward and Caleb Smith were for evacuating Sumter, but they both wished to reinforce Fort Pickens. Bates proposed to strengthen Pickens and Key West and said that the time had come either to evacuate Sumter or relieve it. The rest were unequivocally for a relief expedition. Later that day Lincoln directed the secretaries of War and the Navy to co-operate in preparing an expedition to move by sea as early as April 6. The destination was not indicated in the order, but it was Charleston.[98]

On the same day Seward, intent upon the reinforcement of Fort Pickens, brought Captain M. C. Meigs of the Engineers to Lincoln to discuss an expedition to that place. On March 31 Meigs and Colonel Erasmus D. Keyes, of General Scott's staff, were directed to draw up a plan for the relief of Fort Pickens. They took it to Lincoln who had them take it to Scott to be put into final form and executed. On the next day, April 1, Seward, Meigs, and Lieutenant D. D. Porter of the navy went to the Executive Mansion and after consultation with Lincoln finished the plans for the Pickens expedition. It was to be conducted with such absolute secrecy, lest information leak out to the Confederates, that even the

---

[98] Howard K. Beale (ed.), *The Diary of Edward Bates, 1859-1866*, in American Historical Association, *Annual Report, 1930* (Washington, 1933), IV, 180; Nicolay and Lay (eds.), *Lincoln: Works*, II, 25-28.

secretaries of War and the Navy were to know nothing of it. The orders were signed by the President himself. It was only because the same ship, the *Powhatan*, was selected for both expeditions that the Secretary of the Navy learned of the expedition to the Gulf of Mexico.[99] Energetic preparations began in New York and Brooklyn to collect vessels, men, arms, and provisions for the two expeditions.

In the first days of April came the disquieting returns from the elections in Ohio, Connecticut, and Rhode Island. April 4 proved to be an important day. Early that morning Lincoln seems to have had a mysterious conference with a group of Republican governors, said to be seven or nine in number. Among them were Andrew G. Curtin of Pennsylvania, William Dennison of Ohio, Richard Yates of Illinois, Oliver P. Morton of Indiana, Israel Washburn of Maine, and Austin Blair of Michigan.[100] How did all these governors happen to be in

[99] John T. Morse (ed.), *The Diary of Gideon Welles*, 3 vols. (Boston and New York, 1911), I, 23-25. Hereafter cited as Welles, *Diary*. David D. Porter, in *Incidents and Anecdotes of the Civil War* (New York, 1885), 13-14, tells a lively and rather amusing story of the conference with Lincoln on April 1.
[100] New York *World*, April 5, 1861; New York *Herald*, April 5, 7, 1861; Philadelphia *Enquirer*, April 6, 1861; James Ford Rhodes, *History of the United States Since the Compromise of 1850*, 8 vols. (New York, 1910 edition) III, 346, n. 3. John B. Baldwin, who had an interview with Lincoln later that morning, testified on February 10, 1866, "At the time I was here I saw, and was introduced to, in the President's room, a number of governors of states. It was at the time the nine governors had the talk here with the President." *Report of the Joint Committee on Reconstruction* (39 Cong., 1 Sess., House Report No. 30), 105. As several of these governors were in Washington for three or four days, it is possible that the

Washington at the same time? The newspapers, in so far as they noticed the presence of these gentlemen, assumed that they were looking after patronage; but rumors were soon current that they had gone to demand of the President that he send relief to the garrison at Fort Sumter. This is not improbable since all these men belonged to the aggressive group of Republicans who had been alarmed at the rumors of evacuation and they could hardly have known what Lincoln had already planned. Several questions arise here. If Lincoln was still hesitating, did they bring pressure upon him and force him to a decision? Or did Lincoln allow them to think they were helping him to decide? Or, if the President had not actually summoned them to a conference, did he seize the opportunity to make sure of their powerful support in case the Confederates should show fight? Were mutual pledges of action and support exchanged that morning?

Later that same morning occurred the much-discussed Lincoln-Baldwin interview. On April 2, apparently at the suggestion of Seward, Lincoln had sent Allan B. Magruder, a Virginia Unionist living in Washington, to Richmond to ask G. W. Summers, the leader of the Unionists in the State Convention, to come to see him at once or to send some other representative from that group. Magruder reached Richmond the next day. As Summers could not leave, John B. Baldwin, another leader of the group, was selected; and Baldwin and Magruder were in Washington early on the morning of April 4. They went to Seward who conducted Baldwin to Lincoln at eleven o'clock. Lincoln took Baldwin alone

---

conferences extended over several days, from about April 3 to 6.

into a bedroom, locked the door and exclaimed "You have come too late!" In the conversation which followed, according to Baldwin's statement, the President asked why the Unionists in the Virginia Convention did not adjourn sine die, as the continuance of the session was a standing menace to him. Baldwin replied that if they should so adjourn without having accomplished anything for the security of the state, another convention would certainly be called and it would be more strongly secessionist. He then urged the President to assure peace to the country and strengthen the border-state Unionists by evacuating both Sumter and Pickens and calling upon the whole people to settle their differences in a national convention. Lincoln replied that his supporters would not permit him to withdraw the garrisons. Baldwin then warned him that if a fight started at Fort Sumter, no matter who started it, war would follow and Virginia would go out of the Union in forty-eight hours. Lincoln became greatly excited and exclaimed, "Why was I not told this a week ago? You have come too late!" This is Baldwin's account;[101] but it is substantiated by several other Virginia Unionists, at least to the extent that it was what Baldwin told them when he returned to Richmond the next day.

But John Minor Botts, a violent Virginia Unionist who by invitation talked with Lincoln on the night of April 7, insisted that Lincoln then told him that he had offered to Baldwin to withdraw Anderson's force from Sumter if

---

[101] Baldwin's testimony, *Report of Joint Committee on Reconstruction*, 102-107; J. B. Baldwin, *Interview between President Lincoln and John B. Baldwin, April 4, 1861* (Staunton, VA., 1866).

the Virginia Convention would adjourn sine die, that he would gladly swap a fort for a state; but that Baldwin refused the offer. When Botts offered to take the proposition to Richmond at once Lincoln replied, "Oh, it is too late; the fleet has sailed and I have no means of communicating with it."[102]

Baldwin always denied that Lincoln had made any such proposal as Botts reported. Did Baldwin lie? He seems to have had a much better reputation for accuracy than Botts and his account of his journey to Washington is accurate as far as it can be checked, whereas Botts' story is full of minor inaccuracies.[103] Besides, Baldwin was a sincere Unionist and voted against secession to the last. Why should he have refused Lincoln's offer and failed to report it to his fellow Unionists in Richmond? Did Botts lie about what Lincoln told him? His extreme prejudices and frequently unwarranted statements on other matters would easily bring this conclusion into the range of possibility, were it not for the fact that Lincoln seems to have told much the same story to others. If Lincoln did, then the question whether the President offered to evacuate Sumter at this stage of his plan becomes an issue of veracity between Lincoln and Baldwin, which obviously places the Virginian at a great disadvantage. But let us consider other factors in the situation. Lincoln had just been holding conferences with the militant Republican governors and evidently had come to some agreement

[102] Botts' testimony, *Report of Joint Committee on Reconstruction*, 114-19; John Minor Botts, *The Great Rebellion* (New York, 1866), 194-202.

[103] The most recent and also the most judicial summary of all the evidence is by Henry T. Shanks, *The Secession Movement in Virginia, 1847-1861* (Richmond, 1934), 192-95.

with them, else why should he greet his visitor with the exclamation, repeated later in the conversation, "You have come too late"? Certainly he could not have referred to the final orders to Fox, for those orders were given later that day. And why did he refuse on the night of April 7, if the Botts story is correct, to permit Botts to take his proposition to Richmond, alleging that the fleet had sailed, when in fact none of the vessels left New York until the next night? Is there not some basis for suspecting that Lincoln had not actually made the offer to Baldwin to evacuate Sumter because he was already bound by some sort of agreement with the Republican governors to send the expedition forward; and that later, desiring above all things to leave the impression that he had done everything in his power to avoid a collision, he dropped hints about an offer which had been flatly refused?

During the afternoon of April 4 Lincoln saw Captain Fox, who was to have charge of the Sumter expedition, and told him of his final determination to send relief to Anderson and that notification of the relief expedition would be sent to the Governor of South Carolina before Fox could possibly arrive off Charleston Harbor.[104] Fox hurried back to New York to push his preparations. At some time that same day Lincoln drafted a letter to Major Anderson, which was copied and signed by the Secretary of War, informing him that relief would be sent him.[105]

On the afternoon of April 6 Secretary Welles received a letter from Captain Henry A. Adams of the navy,

---

[104] Crawford, *Genesis of the Civil War*, 404; William E. Smith, *The Francis Preston Blair Family in Politics*, 2 vols. (New York, 1933), II, 12-13.
[105] Nicolay and Hay, *Abraham Lincoln*, IV, 27-28.

stationed off Fort Pickens, explaining that he had not landed the artillery company at the fort in accordance with General Scott's order of March 12 because of controlling orders from the former Secretary of the Navy to respect the truce of February 29, but stating that he was not ready to obey if ordered to land the men. Welles consulted the President and then hurried off Lieutenant John L. Worden with verbal orders to Captain Adams to land the men at once.[106] This incident gave occasion for a strange statement of Lincoln which deserves notice. In his special message to Congress of July 4, he stated that the expedition for the relief of Sumter was first prepared "to be ultimately used or not according to circumstances," and intimated that, if Pickens had been relieved in March, Sumter would have been evacuated, and that it had not been decided to use the expedition until word came that Fort Pickens had not been reinforced in accordance with the order of March 12.[107] The strange thing about this statement is that word was not received from Adams until April 6, while positive orders had been given two days before to Captain Fox to go ahead with his expedition and at the same time Anderson had been notified to expect it. Had Lincoln become confused about the order of these events? It does not seem probable. Or was he, for effect upon public opinion, trying to strengthen the belief that his hand had been forced, that his pacific intentions had been defeated by circumstances?

On April 1 Lincoln had passed the promise through

---

[106] Welles, *Diary*, I, 29-32. Worden reached Captain Adams' ship on April 12 and the men were landed that night, the very day on which the firing began at Sumter.

[107] Richardson (comp.), *Messages and Papers*, VI, 21-22.

Seward and Justice John A. Campbell to the Confederate Commissioners in Washington that he would notify Governor Pickens if any relief expedition should be sent to Fort Sumter.[108] When they learned of it, several members of his cabinet objected to such notification, but Lincoln insisted; he had his own reasons for so doing. The formal notice which he drafted with his own hand, dated April 6, is interesting not only for its careful phrasing but for the evident importance which he attached to it. It was embodied in a letter of instruction to R. S. Chew, an official of the state department who was to be accompanied by Captain Theodore Talbot, directing him to proceed to Charleston where, if he found that Fort Sumter had not been evacuated or attacked and that the flag was still over it, he was to seek an interview with Governor Pickens, read to him the statement and give him a copy of it. If he found the fort evacuated or attacked he was to seek no interview but was to return forthwith. The message to Governor Pickens was in these words:

> "I am directed by the President of the United States to notify you to expect an attempt will be made to supply Fort Sumter with provisions only; and that, if such an attempt be not resisted, no effort to throw in men, arms, or ammunition will be made without further notice, or in case of an attack upon the fort."[109]

---

[108] Connor, *John Archibald Campbell*, 127-28. Lincoln chose to send the notification to the Governor, not the Confederate officers, because he could recognize the former and not the latter.

[109] Nicolay and Hay, *Abraham Lincoln*, IV, 34.

Was the purpose of this message merely to fulfill a promise? Is there no special significance in the fact that Lincoln entrusted the form of it to no one else, but carefully drafted it himself? It is unnecessary to call attention again to the fact that Lincoln was a rare master of the written word, that he had the skill of an artist in so phrasing a sentence that it conveyed precisely the meaning he wished it to convey. He could do more than that: he could make the same sentence say one thing to one person and something entirely different to another and in each case carry the meaning he intended. It is obvious that the message to be read to Governor Pickens was intended less for that official than for General Beauregard and the Confederate government at Montgomery. But it was intended also for the people of the North and of the border states. To the suspicious and apprehensive Confederates it did not merely give information that provisions would be sent to Anderson's garrison — which should be enough to bring about an attempt to take the fort —but it carried a threat that force would be used if the provisions were not allowed to be brought in. It was a direct challenge! How were the Southerners expected to react to this challenge? To Northern readers the same words meant only that the government was taking food to hungry men to whom it was under special obligation. Northern men would see no threat; they would understand only that their government did not propose to use force if it could be avoided. Is it possible that a man of Lincoln's known perspicacity could be blind to the different interpretations which would be placed upon his subtle words in the North and in the South?

The message was not only skillfully phrased, it was most carefully timed. It was read to Governor Pickens in the presence of General Beauregard on the evening of April 8. News of the preparation of some large expedition had been in the newspapers for a week; but as the destination had not been officially divulged, newspaper reporters and correspondents had guessed at many places, chiefly the coast of Texas and revolutionary Santo Domingo. It was not until April 8 that the guessing veered toward Charleston, and not until the next day was any positive information given in the press of the notice to Governor Pickens.[110] The Confederate officials had regarded these preparations at New York with suspicion while conflicting reports came to them from Washington concerning Lincoln's designs about Sumter. The first of Captain Fox's vessels were leaving New York Harbor at the very hour that Chew read the notification to Governor Pickens. The Confederates were given ample time, therefore, to act before the fleet could arrive off Charleston. They did not know that a portion of the vessels which had left New York were really destined not for Charleston but for Fort Pickens at Pensacola. The utmost secrecy was maintained about the Pensacola expedition, thus permitting the Confederates to believe that the whole force was to be concentrated at Charleston.

The tables were now completely turned on the Southerners. Lincoln was well out of his dilemma while they, who had heretofore had the tactical advantage of being able to wait until Anderson must evacuate, were

---

[110] New York *Times*, April 8, 1861; Baltimore *Sun*, April 8, 1861. The Richmond *Examiner* asserted as early as April 6 that the expedition was for the purpose of relieving Sumter.

suddenly faced with a choice of two evils. They must either take the fort before relief could arrive, thus taking the apparent offensive which they had hoped to avoid, or they must stand by quietly and see the fort provisioned. But to allow the provisioning meant not only an indefinite postponement to their possession of the fort which had become as much a symbol to them as it was to Lincoln; to permit it in the face of the threat of force, after all their preparations, would be to make a ridiculous and disgraceful retreat.[111] Nor could they be sure that, if they yielded now in the matter of "Provisions only," they would not soon be served with the "further notice" as a prelude to throwing in "Men, arms, and ammunition." This, then, was the dilemma which they faced as the result of Lincoln's astute strategy.

Events now hurried to the inevitable climax. As soon as President Lincoln's communication was received General Beauregard telegraphed the news to the Confederate secretary of war, L. P. Walker. Walker at once ordered that the Sumter garrison be isolated by stopping its mails and the purchase of provisions in Charleston. On this same day the Confederate commissioners at Washington had received a copy of a memorandum filed in the state department by Seward,

---

[111] Evidently Lincoln did not expect them to retreat, for on April 8 he wrote Governor Curtin of Pennsylvania, one of the recent conferees, "I think the necessity of being *ready* increases. Look to it." From "Lincoln Photostats," Library of Congress; also in Paul M. Angle, *New Letters and Papers of Lincoln* (Boston and New York, 1930), 266. Governor Dennison of Ohio, who was still in Washington, was quoted as promising, on the same date, support to "a vigorous policy." Mt. Vernon (Ohio) *Democratic Banner*, April 16, 1861.

dated March 15, in which the Secretary declined to hold any official intercourse with them. They telegraphed the news to their government and at once, feeling that they had been deceived and knowing that their mission had failed, prepared to leave Washington. Jefferson Davis was thus, on April 8, apprised of two movements by the Federal government which, taken together or singly, looked ominous. On the following day Beauregard seized the mails as they came from Fort Sumter and discovered a letter from Anderson to the war department which disclosed that he had been informed of the coming of Fox's expedition and indicated that the fleet would attempt to force its way into the harbor. This information also was at once communicated to the Montgomery government. On the tenth came the news that the fleet had sailed from New York. Walker then directed Beauregard, if he thought there was no doubt of the authorized character of the notification from Washington (meaning Lincoln's), to demand the evacuation of Fort Sumter and, if it should be refused, "to reduce" the fort. The Davis administration had waited two full days after receiving word of Lincoln's notification before deciding what to do. It is said that Robert Toombs, secretary of state, objected vigorously to attacking the fort. "It is unnecessary; it puts us in the wrong; it is fatal!"[112] If

---

[112] That Toombs protested against the attack seems to be based wholly upon the statement in Pleasant A. Stovall, *The Life of Robert Toombs* (New York, 1892), 226. Stovall cites no source and U. B. Phillips in his *Life of Robert Toombs* (New York, 1913), 234-35, gives no other citation than Stovall. Richard Lathers attributed the same words to Toombs several days before this crisis arose in a letter which he wrote to the New York *Journal of Commerce* from Montgomery. See Alvan

Toombs protested, he was overruled because Davis and the rest believed that Lincoln had already taken the aggressive and they regarded their problem now as a military one. To them it was the simple question whether they should permit the hostile fleet to arrive before they attacked the fort or whether they should take Sumter before they had to fight both fort and fleet.

At two o'clock on the eleventh Beauregard made the demand upon Anderson, who rejected it but added verbally to the officer sent to him that if not battered to pieces, he would be starved out in a few days. When Beauregard reported this remark to Walker, that official informed him that the government did "Not desire needlessly to bombard Fort Sumter" and that if Major Anderson would state when he would evacuate, Beauregard should "avoid the effusion of blood." Evidently the Montgomery officials thought there was still a chance to get the fort peaceably before the fleet could arrive. Had not Lincoln so carefully timed his message with the movement of Fox there might have been no attack. But late in the afternoon of the same day Beauregard received information from a scout boat that the *Harriet Lane*, one of Fox's ships, had been sighted a few miles out of the harbor. It was expected that all the fleet would be at hand by next day. Nevertheless,

---

F. Sanborn, *Reminiscences of Richard Lathers* (New York, 1907), 164-65. Nevertheless, that Toombs was greatly concerned over the dangers in the situation is attested by the Confederate secretary of war, L. P. Walker, who quotes Toombs as saying at the cabinet meeting on April 10, "The firing upon that fort will inaugurate a civil war greater than any the world has yet seen; and I do not feel competent to advise you." Crawford, *The Genesis of the Civil War*, 421.

Beauregard about midnight sent a second message to Anderson, in accordance with Walker's instructions, saying that if he would state the time at which he would evacuate and would agree not to use "your guns against us unless ours should be employed against Fort Sumter, we will abstain from opening fire upon you." To this Anderson replied that he would evacuate by noon on the fifteenth and would in the meantime not open fire upon Beauregard's forces unless compelled to do so by some hostile act "against the fort or the flag it bears, should I not receive prior to that time controlling instructions from my government or additional supplies." This answer was conditional and unsatisfactory for it was clear that, with Fox's fleet arriving, Anderson would not evacuate. Thereupon the two aids who had carried Beauregard's message, in accordance with their instructions from that office, formally notified Anderson — it was now 3:20 in the morning of the twelfth — that fire would be opened upon him in one hour's time.

What followed we all know. The bombardment which began at 4:30 on the morning of April 12 ended in the surrender of Anderson and his garrison during the afternoon of the following day. The three vessels[113] of the

---

[113] These were the *Baltic*, the *Harriet Lane*, and the *Pawnee*. The *Pocahontas* did not arrive until the 13th. It is an interesting question whether the Northern reaction would have been different if the Confederates had ignored Fort Sumter and concentrated their efforts upon trying to keep the fleet from entering the harbor. The fact that their chief naval officer, Captain Henry J. Hartstene, reported on April 10 that the Federals would be able to reach the fort in boats at night and that he had no vessels strong enough to prevent the entrance of the fleet may have determined the Confederates to take the

fleet which lay outside were unable to get into the harbor because of the high seas and the failure of the rest of the fleet — the tugboats and the *Powhatan* — to arrive. Although there were no casualties during the bombardment, the mere news that the attack on the fort had begun swept the entire North into a roaring flame of anger. The "rebels" had fired the first shot; they had chosen to begin war. If there had been any doubt earlier whether the mass of the Northern people would support the administration in suppressing the secessionists, there was none now. Lincoln's strategy had been completely successful. He seized at once the psychological moment for calling out the militia and committing the North to support of the war. This action cost him four of the border slaves states, but he had probably already discounted that loss.

Perhaps the facts thus far enumerated, standing alone, could hardly be conclusive evidence that Lincoln, having decided that there was no other way than war for the salvation of his administration, his party, and the Union, maneuvered the Confederates into firing the first shot in order that they, rather than he, should take the blame of beginning bloodshed. Though subject to that interpretation, they are also subject to the one which he built up so carefully. It there other evidence? No one, surely, would expect to find in any written word of his a confession of the stratagem; for to acknowledge it openly would have been to destroy the very effect he had been at so much pains to produce. There are, it is true, two statements by him to Captain Fox which are at least suggestive. Fox relates that in their conference of April 4

---

fort first. *Official Records*, Ser. I, Vol. I, 299.

the President told him that he had decided to let the expedition go and that a messenger would be sent to the authorities at Charleston before Fox could possibly get there; and when the Captain reminded the President of the short time in which he must organize the expedition and reach the destined point, Lincoln replied, "You will best fulfill your duty to your country by making the attempt." Then, again, in the letter which Lincoln wrote the chagrined Captain on May 1 to console him for the failure of the fleet to enter Charleston Harbor, he said: "You and I both anticipated that the cause of the country would be advanced by making the attempt to provision Fort Sumter, even if it should fail; and it is no small consolation now to feel that our anticipation is justified by the result."[114] Was this statement merely intended to soothe a disappointed commander, or did it contain a hint that the real objective of the expedition was not at all the relief of Sumter?

Lincoln's two secretaries, John G. Nicolay and John Hay, in their long but not impartial account of the Sumter affair come so close to divulging the essence of the stratagem that one cannot but suspect that they knew of it. In one place they say, with reference to Lincoln's solution of this problem of Sumter, "Abstractly it was enough that the Government was in the right. But to make the issue sure, he determined that in addition the rebellion should be put in the wrong." And again, "President Lincoln in deciding the Sumter question had

---

[114] Crawford, *Genesis of the Civil War*, 404; Robert Means Thompson and Richard Wainwright (eds.), *Confidential Correspondence of Gustavus Vasa Fox*, 2 vols. (New York, 1918), I, 43-44; Nicolay and Hay (eds.), *Lincoln: Works*, II, 41.

adopted a simple but effective policy. To use his own words, he determined to 'send bread to Anderson'; if the rebels fired on that, they would not be able to convince the world that he had begun the civil war." And still later, "When he finally gave the order that the fleet should sail he was master of the situation . . . master if the rebels hesitated or repented, because they would thereby forfeit their prestige with the South; master if they persisted, for he would then command a united North."[115]

Perhaps not much weight should be given to the fact that before the expedition reached Charleston his political opponents in the North expressed suspicion of a design to force civil war upon the country in order to save the Republican party from the disaster threatened in the recent elections and that after the fighting began they roundly accused him of having deliberately provoked it by his demonstration against Charleston. And perhaps there is no significance in the further fact that the more aggressive members of his own party had demanded action to save the party and that the administration newspapers began to assert as soon as the fleet sailed that, if war came, the rebels would be the aggressors.[116]

---

[115] Nicolay and Hay, *Abraham Lincoln*, IV, 33, 44, 62.

[116] Predictions, on the one hand, that the "rebels" would soon start a war and charges, on the other, that, to save the Republican party, Lincoln was demonstrating against Charleston in order to force the Southerners to attack Sumter are to be found in administration and antiadministration papers, respectively, during the week before the fort was fired upon. See, for instance, the Columbus (Ohio) *Crisis*, April 4, 1861; New York *Times*, April 8, 10, 1861; Baltimore *Sun*, April 10, 1861. When the news came of the bombardment at Charleston, the Providence *Daily Post*, April 13, 1861, began an editorial entitled "WHY?" with: "We are to have civil war, if

There is evidence much more to the point than any of these things. Stephen A. Douglas, senator from Illinois, died on June 3, 1861. On June 12 the Republican governor of that state, Richard Yates, appointed to the vacancy Orville H. Browning, a prominent lawyer, a former Whig, then an ardent Republican, and for more than twenty years a personal friend of Abraham Lincoln. Browning was one of the group who from the first had favored vigorous measures and had opposed compromise. He was to become the spokesman of the administration in the Senate. On July 2, 1861, Browning arrived in

---

at all, because Abraham Lincoln loves a party better than he loves his country." And after commenting on what seemed to be a sudden change of policy with respect to Sumter, "Why? We think the reader will perceive why. Mr. Lincoln saw an opportunity to inaugurate civil war without appearing in the character of an aggressor. There are men in Fort Sumter, he said, who are nearly out of provisions. They ought to be fed. We will attempt to feed them. Certainly nobody can blame us for that. . . . The secessionists, who are both mad and foolish, will resist us. They will commence civil war. Then I will appeal to the North to aid me in putting down rebellion, and the North must respond. How can it do otherwise? And sure enough, how can we do otherwise?" A photostatic copy of this editorial was furnished me through the kindness of Professor E. M. Coulter of the University of Georgia.

One story that seems to have had some currency was related by Alexander Long, a Democratic congressman from Ohio, in an antiadministration speech before the House on April 8, 1864, to the effect that when Lincoln first heard the news that the Confederates had opened fire on Fort Sumter, he exclaimed, "I knew they would do it!" *Congressional Globe*, 38 Cong., I Sess., 1499 *et seq*. Long's speech aroused much excitement among the Republicans who attempted to expel him from the House on the ground that he was a sympathizer with the rebellion.

Washington to take his seat in the Senate for the special session which had been called to meet on July 4. On the evening of the third he called at the White House to see his old acquaintance. Now Browning for many years had kept a diary, a fact that very probably was unknown to Lincoln since diarists usually conceal this pleasant and useful vice. In the entry for July 3 Browning relates the conversation he had with the President that evening, for after reading the new Senator his special message to Congress, Lincoln laid aside the document and talked. The rest of the entry best be given in Browning's own words:

> He told me that the very first thing placed in his hands after his inauguration was a letter from Majr Anderson announcing the impossibility of defending or relieving Sumter. That he called the cabinet together and consulted Genl Scott — that Scott concurred with Anderson, and the cabinet, with the exception of P M Genl Blair were for evacuating the Fort, and all the troubles and anxieties of his life had not equaled those which intervened between this time and the fall of Sumter. He himself conceived the idea, and proposed sending supplies, without an attempt to reinforce giving notice of the fact to Gov Pickens of S.C. The plan succeeded. They attacked Sumter — it fell, and thus did more service than it otherwise could.[117]

---

[117] Theodore Calvin Pease and James G. Randall (eds.), *The*

This statement, condensed from the words of Lincoln himself by a close friend who wrote them down when he returned that night to his room at "Mrs. Carter's on Capitol Hill," needs no elaboration. It completes the evidence.

It is not difficult to understand how the usually secretive Lincoln, so long surrounded by strangers and criticized by many whom he had expected to be helpful, talking that night for the first time in many months to an old, loyal, and discreet friend, though a friend who had often been somewhat patronizing, for once forgot to be reticent. It must have been an emotional relief to him, with his pride over his consummate strategy bottled up within him for so long, to be able to impress his friend Browning with his success in meeting a perplexing and dangerous situation. He did not suspect that Browning would set it down in a diary.

There is little more to be said. Some of us will be content to find new reason for admiration of Abraham Lincoln in reflecting on this bit of masterful strategy at the very beginning of his long struggle for the preservation of the Union. Some, perhaps, will be reminded of the famous incident of the Ems telegram of which the cynical Bismarck boasted in his memoirs. And some will wonder whether the sense of responsibility for the actual beginning of a frightful war, far more terrible than he could possibly have foreseen in that early April of 1861, may have deepened the melancholy and the charity toward his Southern foemen which that strange man in the White House was to reveal so often before that final

---

Diary of Orville H. Browning, 2 vols. (Springfield, Ill., 1927), I, 475-76.

tragic April of 1865.

The Natural Limits of Slavery Expansion[118]

by

Charles W. Ramsdell

In summary and conclusion: it seems evident that slavery had about reached its zenith by 1860 and must shortly have begun to decline, for the economic forces which had carried it into the region west of the Mississippi had about reached their maximum effectiveness. It could not go forward in any direction and it was losing ground along its northern border.

In the forefront of that group of issues which, for more than a decade before the secession of the cotton states, kept the northern and southern sections of the United States in irritating controversy and a growing sense of enmity, was the question whether the federal government should permit and protect the expansion of slavery into the western territories. If it be granted that this was not at all times the foremost cause of controversy between the

---

[118] Charles W. Ramsdell, "The Natural Limits of Slavery Expansion." *The Southwestern Historical Quarterly* Vol. 33, No. 2 (Oct., 1929), 91-111; originally published in *The Mississippi Valley Historical Review*, Vol. 16, No. 2 (Sept., 1929), 151-171. Article and citation are verbatim.

sections, it must he acknowledged that no other question was the subject of such continuous and widespread interest nor of such acrimonious debate. While behind it lay the larger question whether slavery should be allowed to persist permanently where it already existed, it was this immediate problem of the extension of the institution that gave excitement to the political contests of 1843 to 1845, of 1847 to 1851, and of 1854 to 1860. It was upon this particular issue that a new and powerful sectional party appeared in 1854, that the majority of the Secessionists of the cotton states predicated their action in 1860 and 1861, and it was upon this also that President-elect Lincoln forced the defeat of the compromise measures in the winter of 1860-61. It seems safe to say that had this question been eliminated or settled amicably, there would have been no secession and no Civil War.

The essential points in the controversy over slavery expansion are well known; but in order to focus attention upon the phase of the question here under discussion, it is desirable to cite them again. As stated by the supporters of the Wilmot Proviso and the opponents of the Kansas-Nebraska Bill, it was the question whether the plantation system of agriculture and negro slave labor should be allowed to take possession of the vast western plains, shut out the white home-owning small farmer and the white free laborer, and, by the creation of new slave states, so far increase the political strength of the "slave power" that it would be able to dominate the whole nation in its own interest. As stated by the pro-slavery men, it was the question whether an important and essential southern interest, guaranteed by the federal compact, should be stigmatized by the general government itself and excluded from the territories owned in common by all the states, with the inevitable consequence of so weakening the southern people politically that they would soon no longer be able to defend themselves against hostile and ruinous legislation. This brief explanation does not cover all the ground, but it may suffice for the present purpose. Each party to the controversy considered itself on the defensive and, therefore, to each the issue seemed of vital importance. Neither was willing to surrender anything.

Disregarding the stock arguments — constitutional, economic, social, and what not — advanced by either group, let us examine afresh the real problem involved. Would slavery, if legally permitted to do so, have taken possession of the territories or of any considerable portion of them? There is no question but that our own generation must, if the fears of the anti-expansionists were well founded, sympathize with the opposition to slavery extension. But were their apprehensions well founded? A number of eminent historians, while admitting that slavery could not have flourished on the high arid lands of New Mexico, have either ignored the question with respect to Kansas or have tacitly seemed to assume that the upper plains region would have become a slave section but for the uprising of the people of the free states. They have pointed to various projects for annexations or protectorates to the south of the United States as further evidence of a dangerous program for the extension of the slave power. They have applauded the prophecy of Lincoln, in his "house-divided" speech, that slavery, if not arrested, would extend over the whole country, North as well as South. Despite a lingering disinclination to question Lincoln's infallibility, probably few students of that period today would fully subscribe to that belief. Indeed, many of them have already expressed their disbelief; but so far as I am aware the subject has never been examined comprehensively and the results set down. It is time that such an examination should be made; and, since those more competent have not attempted it, I shall endeavor in this paper to direct attention to the question, even if I throw little new light upon it.

The causes of the expansion of slavery westward from the South Atlantic Coast are now well understood. The industrial revolution and the opening of world markets had continually increased the consumption and demand for raw cotton, while the abundance of fertile and cheap cotton lands in the Gulf States had steadily lured cotton farmers and planters westward. Where large-scale production was possible, the enormous demand for a steady supply of labor had made the use of slaves inevitable, for a sufficient supply of free labor was unprocurable on the frontier. Within one generation, the cotton-growing slave belt had swept across the Gulf region from eastern Georgia to Texas. A parallel movement had carried slaves, though in smaller ratio to whites, into the tobacco and hemp fields of Kentucky, Tennessee, and Missouri. The most powerful factor in the westward movement of slavery was cotton, for the land available for other staples—sugar, hemp, tobacco—was limited, while slave labor was not usually profitable in growing grain. This expansion of the institution was in response to economic stimuli; it had been inspired by no political program nor by any ulterior political purpose. It requires but little acquaintance with the strongly individualistic and unregimented society of that day to see that it would have been extremely difficult, if not impossible, to carry out such an extensive program; nor is there any evidence that such a program existed. There was incentive enough in the desire of the individual slaveowner for the greater profits which he expected in the new lands. The movement would go on as far as suitable cotton lands were to be found or as long as there was a reasonable expectation of profit from slave labor, provided, of course, that no political barrier was encountered. The astonishing rapidity of the advance of the southern frontier prior to 1840 had alarmed the opponents of slavery, who feared that the institution would extend indefinitely into the west. But by 1849-50, when the contest over the principle of the Wilmot Proviso was at its height, the western limits of the cotton-growing region were already approximated; and by the time the new Republican party was formed to check the further expansion of slavery, the westward march of the cotton plantation was evidently slowing down. The northern frontier of cotton production west of the Mississippi had already been established at about the northern line of Arkansas.[119] Only a negligible amount of the staple was being grown in Missouri. West of Arkansas a little cotton was cultivated by the slave holding, civilized Indians; but until the Indian territory should be opened generally to white settlement—a development of which there was no immediate prospect—it could not become a slaveholding region of any importance. The only possibility of a further westward extension of the cotton belt was in Texas. In

---

[119] See charts in *Atlas of American Agriculture* (Washington, 1918), Part V, Sect. A: Cotton, 16-17.

that state alone was the frontier line of cotton and slavery still advancing.

In considering the possibilities of the further extension of slavery, then, it is necessary to examine the situation in Texas in the eighteen-fifties. Though slaves had been introduced into Texas by some of Stephen Austin's colonists, they were not brought in large numbers until after annexation. Before the Texas Revolution, the attitude of the Mexican government and the difficulty of marketing the products of slave labor had checked their introduction; while during the period of the Republic, the uncertainty as to the future of the country, the heavy tariff laid upon Texas cotton by the United States,[120] which in the absence of a direct trade with Europe was virtually the only market for Texas cotton, and the low price of cotton after 1839, had been sufficient in general to restrain the cotton planter from emigrating to the new country. Annexation to the United States and the successful termination of the war with Mexico removed most of these impediments. Thereafter there was no tariff to pay; slave property was safe; land agents offered an abundance of cheap rich lands near enough to the coast and to navigable rivers to permit ready exportation; and the price of cotton was again at a profitable figure. Planters with their slaves poured into the new state in increasing numbers. They settled along the northeastern border, where they had an outlet by way of the Red River, or in the east and southeast along the rivers which flowed into the Gulf. But these rivers were

---

[120] G. P. Garrison, "Texan Diplomatic Correspondence," American Historical Association, *Annual Report,* 1907, I, 522, II, 620; 1908, II, 844, 1215.

not navigable very far from the coast, and the planter who went far into the interior found difficulty in getting his cotton to market. He must either wait upon a rise in the river and depend upon occasional small steamers or the risky method of floating his crop down on rafts; or he must haul it during the wet winter season along nearly impassable pioneer roads and across unbridged streams to Houston or Shreveport, or some other far-off market. The larger his crop, the more time, difficulty, and expense of getting it to market.

Obviously, there was a geographic limit beyond which, under such conditions, the growth of large crops of cotton was unprofitable. Therefore, in the early fifties, the cotton plantations tended to cluster in the river counties in the eastern and southern parts of the state. While the small farmers and stockmen pushed steadily out into the central section of Texas, driving the Indians before them, the cotton plantations and the mass of the slaves lagged far behind. The up-country settlers grew their little crops of grain on some of the finest cotton lands of the world; and they sold their surplus to immigrants and to army posts. Few negroes were to be found on these upland farms, both because the prices demanded for slaves were too high for the farmers to buy them, and because the seasonal character of labor in grain growing rendered the use of slaves unprofitable. Though negro mechanics were in demand and were hired at high wages, the field hand had to be employed fairly steadily throughout the year if his labor was to show a profit. Negroes were even less useful in handling range stock than in farming and were rarely used for that purpose.[121]

[121] J. De Cordova, *Texas: Her Resources and Her Public Men*

Therefore, the extension of the cotton plantation into the interior of Texas had to wait upon the development of a cheaper and more efficient means of transportation. As all attempts to improve the navigation of the shallow, snag-filled rivers failed, it became more and more evident that the only solution of the problem of the interior planter lay in the building of railroads. Throughout the eighteen-fifties, and indeed for two decades after the war, there was a feverish demand for railroads in all parts of the state. The newspapers of the period were full of projects and promises, and scores of railroad companies were organized or promoted. But capital was lacking and the roads were slow in building. Not a single railroad had reached the fertile black-land belt of central Texas by 1860. There can hardly be any question that the cotton plantations ·with their working forces of slaves would have followed the railroads westward until they reached the black-land prairies of central Texas or the semi-arid plains which cover the western half of the state. But would they have followed on into the prairies and the plains ?

It is important to recall that eastern Texas, like the older South Atlantic and Gulf cotton region, is a wooded country, where the essential problem of enclosing fields was easily solved by the rail fence. But in the black-land prairies there was no fencing material, except for a little wood along the creeks; and during the fifties the small fields of the farmers were along these streams. The prairies, generally, were not enclosed and put under the plow until after the introduction of barbed wire in the late

---

(Philadelphia, 1858), 352-53; Frederick Law Olmsted, *A Journey through Texas . . .* (New York, 1860), 440-41.

seventies.[122] Unless the planter had resorted to the expense of shipping rails from eastern Texas, there was no way in which he could have made more use of the prairie lands than the small farmers did. Here, then, in the central black-land prairies, was a temporary barrier to the westward movement of the slave plantation. Beyond it was another barrier that would have been permanently impassable.

Running north and south, just west of the black-land belt, and almost in the geographical center of the state, is a hilly, wooded strip of varying width known as the East and West Cross Timbers, which is prolonged to the south and southwest by the Edwards Plateau. West of the Cross Timbers begins the semi-arid plain which rises to the high, flat table-land of the Staked Plains, or Llano Estacado, in the extreme west and northwest.

Except for a few small cattle ranches, there were almost no settlements in this plains country before 1860; and despite the heavy immigration into Texas after the Civil War, it was not until the eighties that farmers began to penetrate this section.

The history of the agricultural development of the Texas plains region since 1880 affords abundant evidence that it would never have become suitable for plantation slave labor. Let us turn, for a moment, to this later period. The Texas and Pacific Railroad, completed by 1882 and followed by the building of other roads into and across the plains, afforded transportation; and the introduction of

---

[122] Smooth wire was used for fencing to some extent in the fifties, but it was expensive and was ineffective for keeping cattle out of the fields. Barbed wire, invented in 1873, began to be used extensively in Texas about 1878 or 1879.

barbed wire solved the fencing problem. State and railroad lands were offered the settlers at low prices. Farmers began moving into the eastern plains about 1880, but they were driven back again and again by droughts. It took more than twenty years of experimentation and adaptation with wind-mills, dry-farming, and new drought-resisting feed crops for the cotton farmer to conquer the plains.[123] There is little reason to believe that the conquest could have been effected earlier; there is even less basis for belief that the region would ever have been filled with plantations and slaves. For reasons which will be advanced later, it is likely that the institution of slavery would have declined toward extinction in the Old South before the cotton conquest of the plains could have been accomplished, even had there been no Civil War. But if the institution had remained in full vigor elsewhere, it would have been almost impossible to establish the plantation system in this semi-arid section where, in the experimental period, complete losses of crops were so frequent. With so much of his capital tied up in unremunerative laborers whom he must feed and clothe, it is hard to see how any planter could have stayed in that country. Moreover, in the later period the use of improved machinery, especially adapted to the plains, would have made slave labor unnecessary and unbearably expensive. The character of the soil and the infrequency of rainfall have enabled the western

[123] I am under special obligations to my colleague, Professor W. P. Webb, for suggestions concerning the difficulties of the early farmers in western Texas. I have also some first-hand knowledge of the subject, for I lived in the plains region as a boy during a time when the country was almost abandoned because of a prolonged drought.

cotton farmer, since 1900, with the use of this improved machinery to cultivate a far larger acreage in cotton, and other crops as well, than was possible in the older South or in eastern Texas. The result has been the appearance of a high peak in the demand for labor in western Texas in the cotton-picking season. This has called for transient or seasonal labor as in the grain fields—a situation that could not be met by the plantation system of slave labor. During the last twenty-five years this section has become populous and prosperous; but the beginning of its success as a cotton-growing region came fifty years after the Republican party was organized to stop the westward advance of the "cotton barons" and their slaves. It may or may not have any significance that the negro has moved but little farther west in Texas than he was in 1860—he is still a rarity in the plains country although it may be presumed that his labor has been cheaper in freedom than under slavery.

But let us look for a moment at the southwestern border of Texas. In 1860 slavery had stopped more than one hundred and fifty miles short of the Rio Grande. One obvious explanation of this fact is that the slaveowner feared to get too close to the boundary lest his bondmen escape into Mexico. There is no doubt that this fear existed, and that slaves occasionally made their way into that country. But it is worth noting that very little cotton was grown then or is yet grown on that border of Texas, except in the lower valley around Brownsville and along the coast about Corpus Christi. Other crops have proved better adapted to the soil and climate and have paid better.

More significant still is the fact that very few negroes

are found there today, for Mexican labor is cheaper than negro labor now, as it was in the eighteen-fifties. During the decade before secession, Mexican labor was used exclusively south of the Nueces River. After emancipation there was still no movement of negroes into the region where Mexican labor was employed. The disturbances which began in Mexico in 1910 have sent floods of Mexicans across the Rio Grande to labor in the fruit and truck farms of the valley and the cotton fields of south Texas. An interesting result is that the Mexican has steadily pushed the negro out of south Texas and to a considerable degree out of south-central Texas. Wherever the two have come into competition either on the farms or as day laborers in the towns, the Mexican has won. This would seem to show that there was little chance for the institution of African slavery to make headway in the direction of Mexico.

There was another situation which checked the extension of slavery into southwestern Texas. A large area of the most fertile lands had been settled by German immigrants, who had begun coming into that district in the late eighteen-forties. Not only were the Germans opposed to slavery; they were too poor to purchase slaves. They needed labor, as all pioneers do; but their needs were met by the steady inflow of new German immigrants, whose habit it was to hire themselves out until they were able to buy small farms for themselves. The system of agriculture of these industrious and frugal people had no place for the African, whether slave or free. Even today one sees few negroes among the original and typical German settlements. In 1860, east and southeast of San Antonio, these Germans formed a barrier across the front of the slaveholders.

Before turning to the possibilities of slavery extension in other sections, let us consider another question that may be raised by those who still feel that possibly some political advantage was to be gained for the pro-slavery cause in Texas. It had been provided in the joint resolution for the annexation of Texas, in 1845, that as many as four additional states could be formed from the new state, with the consent of Texas, and that such states as should be formed from the territory "south of the line of thirty-six degrees and thirty minutes north latitude, commonly known as the Missouri Compromise line, shall be admitted into the Union with or without slavery, as the people of each state asking admission may desire." It is frequently said that this division, if made, would have had the effect, politically, of an extension of the slavery system through the addition of at least two and possibly eight pro-slavery votes for the South in the United States Senate. Though there was some suggestion of such a division from time to time in other parts of the South before 1860—and sometimes in the North—the sentiment for it in Texas was negligible and it was never seriously contemplated by any considerable group. A strong state pride, always characteristic of the Texans, was against division. There was some sectional feeling between the east and the west, dating from the days of the Republic; and the only agitation of the subject before the war was in 1850 and 1851 when discontent was expressed in eastern Texas over the selection of Austin as the permanent location of the capital. The agitation was frowned upon by the pro-slavery leaders on the ground that separation would result in the creation of a free state in western Texas, which was then overwhelmingly non-slaveholding.[124]

By the provisions of the Compromise of 1850, New Mexico, Utah, and the other territories acquired from Mexico were legally open to slavery. In view of well-known facts, it may hardly seem worth while to discuss the question whether slavery would ever have taken possession of that vast region; but perhaps some of those facts should be set down. The real western frontier of the cotton belt is still in Texas; for though cotton is grown in small quantities in New Mexico, Arizona, and California, in none of these states is the entire yield equal to that of certain single counties in Texas. In none is negro labor used to any appreciable extent, if at all. In New Mexico and Arizona, Mexican labor is cheaper than negro labor, as has been the case ever since the acquisition of the region from Mexico. It was well understood by sensible men, North and South, in 1850 that soil, climate, and native labor would form a perpetual bar to slavery in the vast territory then called New Mexico. Possibly southern California could have sustained slavery, but California had already decided that question for itself, and there was

---

[124] *The Texas Republican* (Marshall), July 7, Aug. 10, Oct. 4, 1849, May 23, 1850; *Texas State Gazette* (Austin), .June 1, 1850, April 12, 1851; *Northern Standard* (Clarksville), March 23, April 27, 1850. When, in 1857, one newspaper, the Jefferson *Herald,* advocated division in order to create more slave states, the Dallas *Herald* opposed the proposition vigorously on the ground that it would do no good, as the whole West and Northwest would be made into free states, and said that the only protection for the South was in forcing a rigid adherence to the Constitution, failing which "the Union must slide." Quoted in *Texas Republican,* Nov. 7, 1857. There seems to have been little other discussion of the subject.

no remote probability that the decision would ever be reversed. As to New Mexico, the census of 1860, ten years after the territory had been thrown open to slavery, showed not a single slave; and this was true, also, of both Colorado and Nevada. Utah, alone of all these territories, was credited with any slaves at all. Surely these results for the ten years when, it is alleged, the slave power was doing its utmost to extend its system into the West, ought to have confuted those who had called down frenzied curses upon the head of Daniel Webster for his Seventh-of -March speech.

At the very time when slavery was reaching its natural and impassable frontiers in Texas, there arose the fateful excitement over the Kansas-Nebraska Bill, or rather over the clause which abrogated the Missouri Compromise and left the determination of the status of slavery in the two territories to their own settlers. Every student of American history knows of the explosion produced in the North by the "Appeal of the Independent Democrats in Congress to the People of the United States," written and circulated by Senator Chase and other members of Congress. This fulmination predicted that the passage of the bill would result in debarring free home-seeking immigrants and laborers from a vast region larger, excluding California, than all the free states, and in converting it into a dreary waste filled with plantations and slaves. It was a remarkably skillful maneuver and it set the North, particularly the Northwest, on fire. But, in all candor, what of the truth of the prophecy? Can anyone who examines the matter objectively today say that there was any probability that slavery as an institution would ever have taken possession of either Kansas or

Nebraska?[125] Certainly cotton could not have been grown in either, for it was not grown in the adjacent part of Missouri. Hemp, and possibly tobacco, might have been grown in a limited portion of eastern Kansas along the Missouri and the lower Kansas rivers; and if no obstacle had been present, undoubtedly a few negroes would have been taken into eastern Kansas. But the infiltration of slaves would have been a slow process.

Apparently there was no expectation, even on the part of the pro-slavery men, that slavery would go into Nebraska. Only a small fraction of the territory was suited to any crops that could be grown with profit by slave labor, and by far the greater portion of Kansas—even of the eastern half that was available for immediate settlement—would have been occupied in a short time, as it was in fact, by a predominantly non-slaveholding and free-soil population. To say that the individual slaveowner would disregard his own economic interest and carry valuable property where it would entail loss merely for the sake of a doubtful political advantage seems a palpable absurdity. Indeed, competent students who have examined this subject have shown that the chief interest of the pro-slavery Missourians in seeking to control the organization of the territorial government was not so much in taking slaves into Kansas as in making sure that

[125] It is clear that Senator Stephen A. Douglas believed that neither Nebraska nor Kansas would ever become a slave state. See quotations from his speeches in Albert J. Beveridge, *Abraham Lincoln, 1809-1858* (Boston, 1928), 108, 193. It is well known now, chiefly through the studies of Professor Frank Hodder, that Douglas' purpose in introducing the bill was to promote the building of a Pacific railroad west from Chicago, not to extend slavery.

no free-soil territory should be organized on their border to endanger their property in western Missouri.[126] They lost in the end, as they were bound to lose. The census of 1860 showed two slaves in Kansas and fifteen in Nebraska. In short, there is good reason to believe that had Douglas' bill passed Congress without protest, and had it been sustained by the people of the free states, slavery could not have taken permanent root in Kansas if the decision were left to the people of the territory itself.

The fierce contest which accompanied and followed the passage of Douglas' Kansas-Nebraska Bill is one of the sad ironies of history. Northern and southern politicians and agitators, backed by excited constituents, threw fuel to the flames of sectional antagonism until the country blazed into a civil war that was the greatest tragedy of the nation. There is no need here to analyze the arguments, constitutional or otherwise, that were employed. Each party to the controversy seemed obsessed by the fear that its own preservation was at stake. The northern anti-slavery men held that a legal sanction of slavery in the territories would result in the extension of the institution and the domination of the free North by the slave power; prospective immigrants in particular f eared that they would never be able to get homes in this new West. Their fears were groundless; but in their excited state of mind they could neither see the facts clearly nor consider them calmly. The slaveholding Southerners, along with other thousands of Southerners who never owned slaves,

---

[126] See James C. Malin, "The Proslavery Background of the Kansas Struggle," *Mississippi Valley Historical Review,* X, 285-305; also H. A. Trexler, *Slavery in Missouri, 1804-1865* (Baltimore, 1914), 185-86.

believed that a victory in Kansas for the anti-slavery forces would not only weaken southern defenses—for they well knew that the South·was on the defensive—but would encourage further attacks until the economic life of the South and "white civilization" were destroyed. Though many of them doubted whether slavery would ever take permanent root in Kansas, they feared to yield a legal precedent which could later be used against them. And so they demanded a right which they could not actively use—the legal right to carry slaves where few would or could be taken. The one side fought rancorously for what it was bound to get without fighting; the other, with equal rancor, contended for what in the nature of things it could never use.

No survey of the possibilities for the expansion of slavery would be complete without giving some consideration to another aspect of the subject—the various proposals for the acquisition of Cuba and Nicaragua, for a protectorate over Mexico, and for the reopening of the African slave trade. These matters can be dealt with briefly, for today the facts are fairly well understood.

The movement for the annexation of Cuba was one of mixed motives.[127] There was the traditional American dislike of Spanish colonial rule, strengthened by a natural sympathy for the Cubans, who were believed to wish independence. There was wide-spread irritation over the difficulty of obtaining from the Spanish government any

---

[127] J. H. Latane, "The Diplomacy of the United States in Regard to Cuba," Am. Hist. Assoc., *Annual Report,* 1897, pp. 232-52. Also, R. G. Caldwell, *The Lopez Expeditions to Cuba, 1848-1851* (Princeton, 1915), 28-38.

redress for indignities perpetrated upon American vessels in Cuban ports and the indifference of Spain to claims for losses sustained by American citizens. Many Americans believed that only the acquisition of the island would terminate our perennial diplomatic troubles with Spain. There was the ever-present desire for territorial expansion, which was by no means peculiar to any section of the country. This ambition was reinforced by an extraordinary confidence in the superiority of American political institutions and the blessings which they would confer upon the annexed peoples. There was also the fear on the part of southern men that British pressure upon Spain would result in the abolition of slavery in Cuba and in some way endanger the institution of slavery in the United States; and this fear was heightened by the knowledge that both Great Britain and France were hostile to American acquisition of the island. A powerful incentive in New Orleans, the hotbed of the filibustering movements, and also in New York, was the hope for a lucrative trade with the island after annexation. There is evidence that some of the planters in the newer cotton belt hoped to get a supply of cheaper slaves from Cuba where the prices were about half what they were in the southern states. Finally, there was the desperate hope of the extreme southern-rights group that, by the admission of Cuba to the Union as a slave state, increased political strength would be added to the defenses of the South.

All these motives were so mixed that it is impossible to assign to each its relative weight. The southern demand for annexation, because of the frankness of the pro-slavery leaders who advocated it and because it was made the point of attack by the anti-slavery group, has been magnified out of its true proportion. Even in the South there was nothing like general approval, by responsible men, of the filibustering enterprises of Lopez and Quitman, for many of those pro-slavery leaders who admitted a desire for the island repudiated the suggestion of forcibly seizing it from Spain.[128] Although both Presidents Pierce and Buchanan pressed offers of purchase upon Spain—or sought to do so—they were unwilling to go further when their offers were coldly rejected. In view of the action of the government in smothering Quitman's filibustering effort in 1854, the general political situation in the United States, and the attitude of Great Britain and France, it must be said that the prospect of acquiring Cuba was, at best, remote.

As to Nicaragua and the frequently asserted dictum that William Walker was but the agent of the slavery expansionists, it is now well enough known that ·walker's enterprise was entirely his own and that he had no intention whatever, if successful, of turning over his private conquest to the United States though he endeavored to use the more fanatical pro-slavery men of the South to further his own designs.[129] In fact, until he broke with Commodore Vanderbilt, he had much closer connection with powerful financial interests in New York

---

[128] See, for instance, a speech of Senator Jefferson Davis before the Democratic State Convention in Jackson, Mississippi, July 6, 1859, in Dunbar Rowland (ed.), *Jefferson Davis Constitutionalist* (Jackson, 1923), IV, 80-81. Also a speech of W. W. Boyce of South Carolina in the House of Representatives, Jan. 15, 1855, in opposition to the annexation of Cuba. Pamphlet in library of University of Texas.
[129] William O. Scroggs, *Filibusters and Financiers . . .* (New York,1916), 49-51, 224-29.

than he had with the Southerners. Had Walker succeeded, those pro-slavery expansionists who had applauded him would most certainly have been sorely disappointed in him. There seems to have been little basis for the fear that Nicaragua would ever have become a field for slavery expansion, or that it could have strengthened in any way the institution of slavery in the southern states. Does the history of the subsequent advance of the United States into the southern islands and Central America induce ironical reflection upon the controversies of the eighteen-fifties?

The filibustering projects against Mexico in the decade of the fifties were of no importance. They were but the feeble continuation of those directed early in the century against the northern provinces of Spain. There is little evidence that any responsible southern leaders cherished the design of seizing additional territory from Mexico for the extension of slavery. They knew too well that it was futile to expect that slaves could be used in the high table-lands or even in the low country where cheaper native labor was already plentiful. It is true that in 1858 Senator Sam Houston of Texas introduced in the Senate a resolution for a protectorate over Mexico. But Houston never showed any interest in the expansion of slavery; and his avowed purpose was to restore peace in Mexico, then distracted by revolutions; to protect the border of the United States; and to enable the Mexican government to pay its debts and satisfy its foreign creditors.[130] His proposal was rejected in the Senate. It was hardly a wise one, but it had nothing to do with slavery. Later in the

[130] *Congressional Globe,* 35 Cong., 1 Sess., 716, 735-37, 1679-82, 2630.

same year, President Buchanan recommended to Congress the establishment of a temporary protectorate over the northern provinces of Mexico for the security of the American border;[131] but it is difficult to read into this suggestion any purpose to expand slavery. Not even a permanent protectorate or annexation could have effected an appreciable expansion of the institution.

The agitation for the re-opening of the African slave trade is an interesting episode. Its proponents were a small group of extremists, mostly Secessionists, whose ostensible object was to cheapen the cost of labor for the small farmer who was too poor to pay the high prices for slaves that prevailed in the fifties. Another argument for re-opening the trade was that cheaper slave labor would enable the institution to extend its frontiers into regions where it was too expensive under existing conditions. Finally, the proponents of the movement insisted that unless the cost of slaves declined, the northern tier of slave states would be drained of their negroes until they themselves became free states, thus imperiling the security of the cotton states. There is some reason to suspect that their leaders designed to stir up the anti-slavery element in the North to greater hostility and to renewed attacks in the hope that the South would be driven into secession, which was the ultimate goal of this faction. These agitators were never able to commit a single state to the project, for not only did the border states condemn it but the majority of the people of the Gulf states also. Even Robert Barnwell Rhett, who was at first inclined to support the program, turned against it

[131] James D. Richardson, *A Compilation of the Messages and Papers of the Presidents,* (Washington, 1897), V, 512-14.

because he saw that it was dividing the state-rights faction and weakening the cause of southern unity. This in itself seems highly significant of the southern attitude.

If the conclusions that have been set forth are sound, by 1860 the institution of slavery had virtually reached its natural frontiers in the west. Beyond Texas and Missouri the way was closed. There was no reasonable ground f or expectation that new lands could be acquired south of the United States into which slaves might be taken. There was, in brief, no further place for it to go. In the cold facts of the situation, there was no longer any basis for excited sectional controversy over slavery extension; but the public mind had so long been concerned with the debate that it could not see that the issue had ceased to have validity. In the existing state of the popular mind, therefore, there was still abundant opportunity for the politician to work to his own ends, to play upon prejudice and passion and fear. Blind leaders of the blind! Sowers of the wind, not seeing how near was the approaching harvest of the whirlwind!

Perhaps this paper should end at this point; but it may be useful to push the inquiry a little farther. If slavery could gain no more political territory, would it be able to hold what it had? Were there not clear indications that its area would soon begin to contract? Were there not even some evidences that a new set of conditions were arising within the South itself which would disintegrate the institution? Here, it must be confessed, one enters the field of speculation, which is always dangerous ground for the historian. But there were certain factors in the situation which can be clearly discerned, and it may serve some purpose to indicate them.

Reference has already been made to the increasingly high prices of slaves in the southwestern states throughout the eighteen-fifties. This price-boom was due in part to good prices for cotton; but though there had always previously been a fairly close correlation between cotton and slave prices, the peculiarity of this situation was that slave prices increased much faster than cotton prices from 1850 to the end of 1860.[132] Probably the explanation lies in the abundance of cheap and fertile cotton lands that were available for planting in Louisiana, Arkansas, and Texas. Cheap lands enabled the planter to expand his plantation and to invest a relatively larger amount of his capital in slaves, and the continued good prices for cotton encouraged this expansion. These good prices for slaves were felt all the way back to the oldest slave states, where slave labor was less profitable, and had the effect of drawing away planters and slaves from Maryland, Virginia, North Carolina, Kentucky, and Missouri to the new Southwest. This movement, to be sure, had been going on for several decades, but now the migration from the old border states was causing alarm among the pro-slavery men. Delaware was only nominally a slave state; Maryland's slave population was diminishing steadily. The ratio of slaves to whites was declining year by year in Virginia, Kentucky, and even in

---

[132] For a general survey of the rise of slave prices for this period, see U. B. Phillips, *American Negro Slavery . . .* (New York, 1918), 370-95, especially the chart opposite 370. Texas newspapers contain frequent accounts of sales of slaves, especially in eastern Texas. The figures run slightly higher than those given by Phillips for New Orleans. See *Texas Republican, passim,* 1852-60. Olmsted, *op. cit.,* 107, 114, quotes prices for the hire of slaves in southwestern Texas.

Missouri. The industrial revolution was reaching into these three states, and promised within less than another generation to reduce the economic interest in planting and slaveholding, as already in Maryland, to very small proportions.

The pro-slavery leaders in Virginia and Maryland endeavored to arrest this change by improving the condition of the planter. They renewed their efforts for a direct trade with Europe, and further stimulated interest in agricultural reforms.[133] As already seen, the proponents of the revival of the African slave trade argued that cheaper slave labor in the lower South was necessary to prevent the border states from ultimately becoming free-soil. Though agricultural reform made headway, the other remedies failed to materialize; and the slow but constant transformation of the Atlantic border region proceeded. The greatest impediments were in the reluctance of the families of the old states, where slavery was strongly patriarchal, to part with their family servants, and in the social prestige which attached to the possession of an ample retinue of servants. It was evident, however, that the exodus would go on until the lure of the Southwest lost its force.

As long as there was an abundance of cheap and fertile cotton lands, as there was in Texas, and the prices of cotton remained good, there would be a heavy demand for labor on the new plantations. As far as fresh lands

---

[133] An editorial in the *Cotton Plant and Southern Advertiser* (Baltimore), April 24, 1852, discusses this situation at length. On the reduction of slave forces on Virginia plantations as a feature of agricultural reform, see A. O. Craven, *Soil Exhaustion as a Factor in the Agricultural History of Virginia and Maryland, 1606-1860* (Urbana, 1925), 145, 158, 161.

were concerned, this condition would last for some time, for the supply of lands in Texas alone was enormous. But at the end of the decade, there were unmistakable signs that a sharp decline in cotton prices and planting profits was close at hand. The production of cotton had increased slowly, with some fluctuations, from 1848 to 1857, and the price varied from about ten cents to over thirteen cents a pound on the New York market. But a rapid increase in production began in 1858 and the price declined. The crop of 1860 was twice that of 1850. Probably the increase in production was due in part to the rapid building of railroads throughout the South toward the end of the decade, which brought new lands within reach of markets and increased the cotton acreage; but part of the increase was due to the new fields in Texas. There was every indication of increased production and lower price levels for the future, even if large allowance be made for poor-crop years. There was small chance of reducing the acreage, for the cotton planter could not easily change to another crop. Had not the war intervened, there is every reason to believe that there would have been a continuous overproduction and very low prices throughout the sixties and seventies.

What would have happened then when the new lands of the Southwest had come into full production and the price of cotton had sunk to the point at which it could not be grown with profit on the millions of acres of poorer soils in the older sections? The replenishment of the soil would not have solved the problem for it would only have resulted in the production of more cotton. Even on the better lands the margin of profit would have declined.[134]

---

[134] Improved farm machinery, which was already beginning to

Prices of slaves must have dropped then, even in the Southwest; importation from the border states would have fallen off; thousands of slaves would have become not only unprofitable but a heavy burden, the market for them gone. Those who are familiar with the history of cotton farming, cotton prices, and the depletion of the cotton lands since the Civil War will agree that this is no fanciful picture.

What would have been the effect of this upon the slaveowner's attitude toward emancipation? No preachments about the sacredness of the institution and of constitutional guarantees would have compensated him for the dwindling values of his lands and slaves and the increasing burden of his debts. It should not be forgotten that the final formulation and acceptance of the so-called "pro-slavery philosophy" belonged to a time when slaveowners, in general, were prosperous. With prosperity gone and slaves an increasingly unprofitable burden, year after year, can there be any doubt that thousands of slave-owners would have sought for some means of relief How they might have solved the problem of getting out from under the burden without entire loss of the capital invested in their working force, it is hard to say; but that they would have changed their attitude to-

---

be introduced on the plantations, would certainly have lowered the cost of crop production; but it must have resulted both in an increase of acreage—thus further reducing prices by increasing the total yield—and the replacement of slave labor to some extent by machinery and the reduction by so much of the demand for and the value of slaves. In fact, there is strong reason to believe that the introduction of labor-saving agricultural machinery would have done much to destroy chattel slavery.

ward the institution seems inevitable.

There was one difficulty about the problem of emancipation that has been little understood in the North, one that the Abolitionist refused to admit. It was the question of what to do with the freed negro. Could he take care of himself without becoming a public charge and a social danger? Would it not be necessary to get rid of the slave and the negro at the same time? But to get rid of the negro was manifestly impossible. Should he not then remain under some form of control both in his own interest and in the interest of the larger social order? There is some evidence that this problem was actually being worked out in those older states which had a large population of free negroes. In Virginia and Maryland, where the number of slaves on the plantation had been reduced in the interest of economy as improved farming machinery came into use, free negroes were coming to be relied upon when extra or seasonable labor was required.[135] Though it is impossible to say how far this practice would have gone in substituting free-negro labor for slave labor, it would inevitably have accustomed increasing numbers of employers to the use of free negroes and have weakened by so much the economic interest in slavery. The cost of rearing a slave to the

---

[135] I am indebted to Professor A. O. Craven, of the University of Chicago, for calling my attention to this development; but he should not be held responsible for my conclusions. For evidence of the use of free negroes on one plantation, see J. S. Bassett (ed.), "The Westover Journal of John A. Selden," *Smith College Studies in History* (Northampton, 1921), VI, 270, 296-98, 316. For a more general account of the use of free negroes in Virginia, see John H. Russell, *The Free Negro in Virginia, 1619-1865* (Baltimore, 1913), 146-55.

working age was considerable, and it is well within the probabilities that, in an era of over-stocked plantations and low cotton prices, the planter would have found that he was rearing slaves, as well as growing cotton, at a loss. New codes for the control of the free negroes might easily, in the course of time, have removed the greatest objection on the part of the non-slaveowners to emancipation.

In summary and conclusion: it seems evident that slavery had about reached its zenith by 1860 and must shortly have begun to decline, for the economic forces which had carried it into the region west of the Mississippi had about reached their maximum effectiveness. It could not go forward in any direction and it was losing ground along its northern border. A cumbersome and expensive system, it could show profits only as long as it could find plenty of rich land to cultivate and the world would take the product of its crude labor at a good price. It had reached its limits in both profits and lands. The free farmers in the North who dreaded its further spread had nothing to fear. Even those who wished it destroyed had only to wait a little while—perhaps a generation, probably less. It was summarily destroyed at a frightful cost to the whole country and one third of the nation was impoverished for forty years. One is tempted at this point to reflections upon what has long passed for statesmanship on both sides of that long dead issue. But I have not the heart to indulge them.

The Changing Interpretation of the Civil War[136]

by

Charles W. Ramsdell

Making all necessary allowances for our inability to weigh accurately the imponderables in the history of a great people, can we say with conviction that this war accomplished anything of lasting good that could not and would not have been won by the peaceful processes of social evolution? Is there not ground for the tragic conclusion that it accomplished little which was not otherwise attainable?

No other event in American history has produced such a flood of controversial historical literature as has the Civil War.[137] The reasons for this are plain. The very complexity of the interwoven social, economic, religious, psychological, and political factors, some of them very obscure or elusive, all of them difficult of quantitative analysis, has baffled even the most impartial

[136] Charles W. Ramsdell, "The Changing Interpretation of the Civil War." *The Journal of Southern History* Vol. 3, No. 1 (Feb., 1937), 3-27. Article and citation are verbatim.

[137] This paper was read as the presidential address before the Southern Historical Association at Nashville, Tennessee, November 20, 1936.

investigators. No other event in our history caused such widespread suffering or aroused such partisan feeling. Finally, since the conflict was primarily sectional, mass opinion in each section, reinforced by common memories and prejudices, hardened into a tradition which was all but impervious to criticism.

Contemporary writers, inevitably partisan, explained the steps which led up to hostilities in the terms of the party conflicts of thirty years. Since the belief was common in the North that the secession leaders had attempted to break up the Union because they could not control it, there was little difficulty in joining to this thesis the idea that they had sought control in order to force the institution of slavery not only into the territories but also into the free states. Secession, it was asserted, was the result of a treasonable conspiracy, conceived long before and aided and abetted by Northern "doughfaces" like President Buchanan. The settlement of Texas by slaveholders, the Revolution by which unscrupulous men tore that vast area from Mexico in order to annex it to the United States, the War with Mexico for more slave territory, the Fugitive Slave Law of 1850, the Kansas-Nebraska Act, the judicial conspiracy of the Dred Scott case and, finally, the effort to break up the Union—all (and more) were but links in the chain which bound the slave-owning aristocracy to parricidal treason. Most of these charges had been the peculiar contributions of the abolitionists and as such had been laughed at by conservative Northerners for years; but under stress of the war psychosis they quickly came to be accepted by the majority of the people. Then the final step was to make this the official version of the origins of the desolating conflict.

In the South the explanation was equally simple and fully as veracious. Northern manufacturers and capitalists had joined with fanatical abolitionists to overthrow the constitutional rights of the Southern states, the one in order to subject Southern agriculture to heavy burdens for the promotion of Northern wealth, the other to break down the beneficent Southern social-racial system and, by inaugurating a war of races, destroy white civilization and elevate the African to a position for which he was not fitted. The Southern states, exercising their sovereign rights, had withdrawn from fellowship with the free states in order to protect their people from destruction by a hostile sectional majority. For them it was a war of defense against wicked aggression and threatened subjugation. Thus, in the main, ran the arguments.

For the most part the earliest histories followed these assumptions. Horace Greeley, writing the preface to his *American Conflict* in April, 1864, while men were fighting and dying all the way from northern Virginia to the borders of Texas, could see no cause for this carnage but the efforts of arrogant slaveholders to destroy a government which they could not control. Even the philosophical scientist, Dr. John W. Draper, whose three-volume work[138] was begun in the midst of the war and who endeavored to get at the fundamental differences between the sections by a study of geographic and climatic influences, could see nothing in the immediate causes of the dreadful scourge but the tyranny of the slave power and the treasonable ambitions of the Southern leaders. And so it was with a host of lesser men. One able

---

[138] *The Civil War in America,* 3 vols. (New York, 1867-1870).

Northern writer, George Lunt, whose *Origins of the Late War* appeared in 1866, held that "slavery, though made an occasion, was not in reality the cause of the war," but that Northern politicians had made use of the slavery issue as an avenue to power and had forced war upon the South as a means of maintaining their control. But Lunt, a conservative Massachusetts Whig who had turned Democrat and had been a consistent opponent of the abolitionists, was a lone voice in New England. The most prolific wartime Southern writer, Edward A. Pollard, in his *First Year of the War*[139] declared that the conflict had been brought about by unscrupulous Northern politicians and business men who had consolidated the numerical majority of their own section on the pretext of staying the advance of slavery but with the real design of destroying the constitutional rights of the Southern states and subverting the Constitution itself in order both to seize upon supreme power and to rob their Southern opponents of their property. In short, he turned back upon the free state leaders the same accusations that they had made against the Southern slaveowners. In his more famous postwar book, *The Lost Cause,*[140] Pollard softened his tone but in substance reiterated the charges.

As the war years receded, books and articles dealing with various aspects of the great conflict flowed from the presses in a steady stream. While most of them were concerned with the story of military operations, a considerable number continued the controversy over the responsibility for the waste of life and property. Time does not permit the mention of more than a few of these

---

[139] Second edition, revised (Richmond, 1862).

[140] New York, 1866.

contributions to the wordy warfare.

In 1868 Alexander H. Stephens of Georgia, a Unionist on grounds of expediency until his state seceded, published the first volume of his *Constitutional View of the War between the States* and followed it with the second volume in 1870. Stephens defended the right of secession on the basic theory of the absolute sovereignty of each individual state, the doctrine first clearly set forth by John C. Calhoun. In 1872 Henry Wilson, abolitionist senator from Massachusetts and later vice-president of the United States, published the first of his three volumes on the *History of the Rise and Fall of the Slave Power in America,*[141] the title of which sufficiently indicates the argument. In 1881 Jefferson Davis finished *The Rise and Fall of the Confederate Government,* a defense of the Southern cause and of his administration of the Confederacy. Like Stephens, Davis held that slavery was merely incidental to and not the true cause of the war. The Southern states had only exercised the sovereign right of self-determination in withdrawing from a Union in which they could no longer expect protection of their rights under the Constitution, and war had then been forced upon an independent Southern people by an aggressive and imperialistic government at Washington. The Davis-Stephens argument was legalistic and therefore limited, but it was so strongly stated that it appealed to Southern readers who were seeking for some fundamental principle in the line of the Southern tradition upon which to justify the cause for which they had suffered so much. In the middle eighties two eminent Republicans, James G. Blaine and John A. Logan, gave to

---

[141] Boston, 1872-1877.

the public political memoirs[142] which, while adding nothing of importance to the nationalist dialectics, were widely read and served to strengthen the familiar tenets of their party and section.

Meanwhile, no less effective work in the formulation of the rival traditions was being done by thousands of men of lesser prominence—editors, politicians, preachers, teachers, and platform lecturers. In the North the "Union-savers" and the abolitionists had joined forces early in the war. The abolitionists, strong in the churches, had been able to add religious and moral sanctions to the cause of unionism, thereby adding immeasurably to the popularity of their interpretation of the conflict. In fixing the war guilt upon the secessionists and especially upon the "slaveholding aristocracy," the conviction that religion and morality were on the side of the victors induced a pleasing sense of righteousness; while the evidence that through God's will the nation most favored by Him of all this modern world had been preserved in its territorial integrity, and the most conspicuous outpost of Satan's dominion had been eliminated from America, was a crowning satisfaction. (The equally religious Southerner, accustomed likewise to rely upon the workings of the Divine Will, had some difficulty in adjusting himself to this mysterious manifestation, except when he agreed with the pious old North Carolinian that there had been "a temporary interruption of the workings of Providence.") But the growing Northern tradition gathered strength from other advantages. The Northeast

---

[142] James G. Blaine, *Twenty Years of Congress,* 2 vols. (Norwich, Conn., 1884-1886); John A. Logan, *The Great Conspiracy* (New York, 1886).

especially enjoyed an amazing prosperity during the early postwar years, and eager business men, having quickly learned how useful government could be to business, looked back to the war as the beginning of a better and brighter day. Thousands of others, who had shared neither in the righteous exaltation nor in the new profits, settled down into the opinions of their neighbors. As society adjusts itself to revolutionary change and proceeds to build or extend its institutional edifice upon the new plan, it displays an irresistible impulse to accept and justify the established order. Success justifies itself; in the long run the victor is always right. Again, the literary dominance of New England, where for years most of the histories were written, enabled "the New England point of view" to permeate the thinking of the greater part of the country. Northern textbooks in use throughout the nation fixed ever more firmly in the popular mind the nationalist and antislavery interpretation of the causes and character of the war.

The Southern cause was as much on the defensive in this battle of interpretations as ever the Confederacy had been on bloodier fields. While many families doubtless resented the loss of their slaves, most Southerners professed to be entirely satisfied that the peculiar institution was gone and insisted that they had gone to war to protect their homes from invasion, not to save the property of the slaveholders.[143] Those who were in public

[143] This statement was literally correct, since the Confederate call to arms in 1861 was to repel invasion, but it ignores the slavery issue as a cause of secession while it indicates that after the war the Southerners were more responsive to the attitude of other peoples toward slavery than they had been before the war.

life and were looked upon as the spokesmen of the stricken South saw that it was essential to the welfare of their people, in both business and politics, that reconciliation between the sections be effected as quickly as possible and that the Democratic party should be able to recover its strength in the dominant North. Therefore, they had powerful motives for accepting the results of the war without further recrimination and for stilling controversy by saying as little as possible about the causes. This last consideration, however, had little weight with the masses of the people who saw no improvement in their condition by reason of the triumph of the North. While conceding that secession had proved a mistake in expediency, they stubbornly insisted that it had been justified in principle and by the dangers which had threatened them in 1861.

Long before the survivors of the Civil War had finished giving the public their versions of the great convulsion, a new group of historians had begun to attract attention. In 1876, 1880, and 1883 appeared the initial volumes of the series projected by Hermann von Holst,[144] James Schouler,[145] and John B. McMaster,[146] respectively, who had severally undertaken to explain the history of the country from the Revolution to the Civil War. That their

---

[144] *The Constitutional and Political History of the United States,* 8 vols. (Chicago, 1876-1892), I. This volume was Erst published in Germany in 1873 under the title *Verfassung und Demokratie der Vereinigten Staaten* while von Holst was teaching in the new University of Strassburg.

[145] *History of the United States under the Constitution,* 7 vols. (New York, 1880-1890), I.

[146] *A History of the People of the United States from the Revolution to the Civil War,* 8 vols. (New York, 1883-1892), I.

eyes were fixed upon that struggle as the climax of the story attests their belief in its overwhelming significance. Von Holst and Schouler, holding to the older conception of history as past politics, confined their narratives almost exclusively to political contests and constitutional questions and used as their sources the published writings of the early statesmen, governmental documents, Federal and state, and the debates in Congress. McMaster, whose interest was in social history, added the files of old newspapers to his sources. It would be unfair to these men to reproach them for the narrow range of their source materials, since they had access to no such vast collections as are available to the historical worker today. But in other respects the limitations of Schouler and von Holst are clear enough to anyone who looks into their nearly forgotten volumes. James Schouler was a New England lawyer, thoroughly imbued with the nationalist-antislavery conception of the background of the secession movement and wholly incapable of understanding the point of view of any section other than New England. Von Holst, a German scholar who had suffered much for liberal principles before he came to New York as a penniless immigrant in 1867, looked upon slavery with horror as the very embodiment of evil and upon the proslavery Southern leaders as wicked men. The outbreak of the war in 1861 was the consequence of a long-laid and carefully executed plot of the "arrogant slavocracy." Von Holst had accepted the antislavery tradition *in toto* and had added some embellishments of his own. Both he and Schouler had relied chiefly upon the debates in Congress and on the political hustings and in them they found the Southern arguments already answered to their satisfaction. Although McMaster ranged much farther afield, he was content to set down both facts and arguments as he found them with little attempt at criticism or analysis. That he, too, was thoroughly indoctrinated with the nationalist tradition is shown by the fact that, although his eighth volume on the decade of the fifties did not appear until 1913, it still reflected the older point of view.

In 1890 John G. Nicolay and John Hay completed their *Abraham Lincoln: A History* in ten stately volumes. This work not only contributed powerfully to the growing Lincoln legend, which was ultimately to displace Washington for the "rail-splitter" as the American folk hero, but also gave additional sanction to the nationalist and antislavery interpretation of the war by linking it with the apotheosis of the great war president. In 1892, the year in which von Holst's last volume came from the press, appeared the first two volumes of James Ford Rhodes covering the ten years from the Compromise of 1850 to the election of Abraham Lincoln to the presidency. In 1895 came the third volume which carried the story to the spring of 1862. Rhodes had a broader and sounder conception of the subject matter of history than either Schouler or von Holst and he was far more judicial in handling controversial questions. But he was hampered by an initial lack of understanding of the South and by the generally anti-Southern character of his sources, and he was also clearly influenced by the traditional attitudes of his native Western Reserve district of Ohio and of his later home in Boston. Nevertheless, while he ran true to form in holding that slavery was the sole cause of secession and, therefore, of the war, he made some advance toward middle ground by testifying to the high personal character of Southern leaders and by rejecting the theory that secession was the fruit of a "conspiracy" of the Southern senators in Washington.[147] His lucid and attractive style and his authoritative manner gave his work great popularity and influence.

In 1897 was published *The Middle Period, 1817-1858,* the work of John W. Burgess, Dean of the Faculty of Political Science of Columbia University and a great figure in the academic world of political science and constitutional history. This book was followed in 1901 by his two volumes entitled *The Civil War and the Constitution.* Burgess was a Tennessean who had served in the Union army and whose nationalist proclivities had been strengthened by study in the universities of Germany. Now a thoroughgoing nationalist, dogmatic in opinion and strongly prone to regard every political issue from the standpoint of what he habitually called "the correct principles of political science," he was nevertheless a close student of American constitutional and political history as it was revealed in the arid pages of the *Congressional Globe.* In the preface to *The Middle Period* Burgess defined his attitude very clearly when he said that the history of the United States should be written by a Northerner and "from the northern point of view . . . because the northern point of view is, in the

[147] While this conspiracy theory was of Northern origin it had been given support by the Virginian Edward A. Pollard in his *Life of Jefferson Davis with a Secret History of the Southern Confederacy* . . . (Philadelphia, 1869), 44 *et seq.* Pollard's motive seems to have been to discredit Davis whom he had attacked unceasingly both during the war and afterwards. This *Life* was a scathing assault upon the Confederate president.

main, the correct view" and that, while sincerity must be allowed the Southern people and their leaders, "not one scintilla of justification for secession and rebellion must be expected. The South must acknowledge its error as well as its defeat." But Dean Burgess did not uphold all of the Northern tradition. For instance, he did not admire William Lloyd Garrison or John Brown; he did not regard the settlement of Texas or its revolt against Mexico as a proslavery plot nor the War with Mexico as proslavery aggression. But he held that the policy of the Southern Democrats with respect to slavery in the territories was aggressive, not defensive; that the theory of state sovereignty was unjustifiable either by the Constitution or by "sound political science"; and that the Southern people and their leaders, by refusing obedience to the Federal government, became responsible for the war. His elaborate analysis of the constitutional aspects of the sectional controversies was the most powerful answer yet made to the state sovereignty arguments of Stephens and Davis and made a deep impression on his contemporaries.

We must pass over the contributions of scores of less conspicuous writers of the years before 1900. It is enough to say that at the end of the century the historical scholars, with few exceptions, agreed that the Civil War had been the overshadowing event of all American history and that most of them accepted the orthodox Northern version of the causes and character of that conflict. No convincing presentation of the Southern cause had caught the popular attention for twenty years. All the great histories had been written in the North. The old antagonisms had died down as the North became more and more absorbed with the problems of its expanding economic life, as the once-desolated South began to feel the thrills of returning prosperity and the sons of Union and Confederate veterans rallied together under the flag in the brief war with Spain. While tradition lingers long among the common folk, there were signs that among those of the South the old defensive traditions were slowly disintegrating or changing form. Left without learned assistance in replying to the Northern historians, affected by the nostalgic reminiscences of the aged for the "good old days before the war," they turned to the romances of Thomas Nelson Page and other Southern novelists and recreated the Old South for themselves in terms of moonlight and roses, tall white-pillared porches, minuets, mint juleps, and happy darkies frolicking in the "quarters."

But the historians were not through with this question. By 1900 the growth of the great graduate schools had reached the point at which the historical seminars were beginning to force the complete rewriting of American history. Doctoral dissertations and the flood of other monographs inspired by great teachers began to attack directly and indirectly the assumptions of the older historians both as to the forces which had influenced our history and its traditional interpretations. Frederick J. Turner at the University of Wisconsin and William A. Dunning at Columbia had begun the seminars which were to have such revolutionary consequences upon the interpretation of the whole nine-teenth century in America. Turner set his students to work upon the growth and interrelations of the varied geographical provinces of the United States and he neglected no aspect of the sectional scene—social, economic, religious, political, psychological, or topographical. Dunning first directed his students to the study of Reconstruction and then led them skillfully back through the war into the antebellum situation. McMaster at the University of Pennsylvania and Edward Channing at Harvard set dozens of graduate students to work searching for new light on problems they had encountered in their own notable volumes. At Johns Hopkins, Yale, Chicago, Michigan, and other universities young men and women were being trained in the techniques of historical investigation and writing and were being introduced to profitable fields for research. It was inevitable that many of these youthful enthusiasts who turned eagerly to digging out new material in this fertile and unworked field should be Southerners with a consuming desire to study the history of their own section. They had been sufficiently well trained to appreciate the necessity for an objective attitude, but doubtless many of them along with the thrill of discovery found a keen pleasure in overturning the theories and assumptions of von Holst, Schouler, Rossiter Johnson, and Rhodes. Some of the most important contributions in this new activity came from students of Northern birth who found fascination in Southern history as well as in Northern. They searched through dusty and forgotten official archives, examined old files of long neglected newspapers, and unearthed hitherto unknown collections of private papers. It is not surprising that they found many of the assumptions of the elder historians defective through lack of accurate or sufficient information.

Time does not permit the mention of all who made important contributions to the revision of this phase of

American history, but a few must be noted. In his studies of the plantation and the regime of slave management, Ulrich B. Phillips[148] thoroughly exploded the abolitionist charge that the slave was systematically or usually overworked or other-wise treated with brutality. His findings, based upon the examination of countless plantation records and related documents and thoroughly objective (for he neither defended nor condemned the system), are so conclusive that it would be impossible for any reputable historian today to describe the institution as did von Holst or Rhodes. Phillips also made it clear that the mass of the Southern people were far less concerned about property rights in slaves (since three fourths of them owned none) than in the underlying racial-social problem involved in emancipation. To put it in another way, they opposed the abolition program because they feared it would ruin the South as "a white man's country." A Northern student, Arthur C. Cole, in his *Whig Party in the South*[149] proved that, contrary to Northern belief, the slaveholders were mostly Whigs who for the sake of party unity as well as for safety discountenanced agitation over abolition, generally opposed aggressive tactics for the extension of slavery, and were rather consistently Union men who flouted the theory of state sovereignty although devoted to the constitutional rights of the states. It was the Democrats, few of whom owned slaves, who were the more aggressive party. Eugene C. Barker's authoritative studies in the field

[148] *American Negro Slavery* (New York and London, 1918); *Life and Labor in the Old South* (Boston, 1929); "The Central Theme of Southern History," in *American Historical Review* (New York, 1895-), XXXIV (1929), 30-43; etc.
[149] Washington, 1914.

of Texas colonization[150] made it clear that slavery extension had nothing to do with the Anglo-American colonization of Texas or the Revolution against Mexico. Justin H. Smith, a Northern historian, after elaborate study came to the conclusion that Mexico, not the United States, was responsible for the war in 1846.[151] Chauncey S. Boucher, of Northern birth, in a notable article[152] pointed out the fallacies in the theory that a united and "aggressive slavocracy" had brought about the annexation of Texas, the war with Mexico, and the seizure of a large portion of that distracted country for the purpose of creating more slave states. Boucher held, as did other investigators, that the Southerners were on the defensive instead of the offensive throughout the whole slavery controversy. Elaborating this point, Jesse T. Carpenter[153] showed how Southern men, always in the minority and conscious of danger to their interests, had erected one defense after another under the Constitution and had finally taken refuge in independence as a last resort when all the others had broken down. Dwight L. Dumond in a careful analysis of the secession movement[154] showed

---

[150] Especially "The Influence of Slavery on the Colonization of Texas," in *Mississippi Valley Historical Review* (Cedar Rapids, 1914-), XI (1925), 3-36; *The Life of Stephen F. Austin* (Nashville, 1925); *Mexico and Texas, 1821-1835* (Dallas, 1928).

[151] *The War with Mexico,* 2 vols. (New York, 1919).

[152] "In Re That Aggressive Slavocracy," in *Mississippi Valley Historical Review,* VIII (1922), 13-79.

[153] *The South as a Conscious Minority, 1789-1861* (New York, 1930).

[154] *The Secession Movement, 1860-1861* (New York, 1931); *Southern Editorials on Secession* (New York and London, 1931).

that, from evidence then available, the Southern leaders had ample reason in 1860 to believe that the South was in real danger from an increasingly hostile majority in the free states, but that even in the face of this situation they had great difficulty in uniting on any course of action. Other investigators have directed their attention to the economic life of the Old South with the result that some of our earlier ideas about that subject have had to be revised radically. But as many of these studies were not directly concerned with the causes of the war, only one will be mentioned here. Robert R. Russel's *Economic Aspects of Southern Sectionalism*[155] shows how Southern discontent over the losing battle which Southern agriculture was waging with the rising Northern industry and capital induced attempts to develop similar industries in the South and contributed to the idea of political independence.

Meanwhile, there has been considerable revision of earlier beliefs about the relation of certain groups in the North to the sectional controversy. One of the most interesting is the rehabilitation of Stephen A. Douglas who had been disparaged in the mistaken idea that this belittling added to the stature of Lincoln. Several writers have contributed to this new understanding of the Little Giant—among them Allen Johnson, Albert J. Beveridge, William O. Lynch, Frank H. Hodder, and George Fort Milton—but space permits the mention of the work of but one. The late Professor Hodder showed conclusively[156]

---

[155] Urbana, Ill., 1923.

[156] "The Genesis of the Kansas-Nebraska Act," in *Proceedings of the State Historical Society of Wisconsin* (Madison, 1874-), Sixtieth Annual Meeting (1912), 69-87; "The Railroad

that Douglas' introduction of the Nebraska bill in 1854 was not a part of any bargain with the South and that his acceptance of the amendment to repeal the Missouri Compromise was not a bid for the presidency. Hodder also proved that the famous *obiter dictum* of the Supreme Court majority in the Dred Scott case was not the result of collusion with the proslavery leaders but was, in a measure, forced by the two minority justices.[157]

Thus, one by one, these old partisan charges of conspiracy and corruption, once accepted by credulous historians as proven facts, are deleted from the page of History. In this connection it may be recalled that abundant evidence has shown that the entire quarrel over the question whether slavery should be permitted in the territories had no basis in fact, but was a mere jockeying for strategic advantage, since no possibility existed that the institution could maintain itself in those regions. Indeed, it becomes more and more apparent that certain of the statesmen of that day were more concerned with immediate political prospects than with the eternal principles of truth which they professed to serve.

Perhaps some of the most fruitful of the newer studies have been those which have re-examined the growth of the antislavery movement and its springs of action. For instance, the recent book of Gilbert H. Barnes, *The Antislavery Impulse, 1830-1844,*[158] and the related two volumes of *The Weld-Grimke Letters,*[159] edited by

---

Background of the Kansas-Nebraska Act," in *Mississippi Valley Historical Review,* XII (1926), 3-22.
[157] "Some Phases of the Dred Scott Case," in *Mississippi Valley Historical Review,* XVI (1930), 3-22.
[158] New York and London, 1933.
[159] Two vols., New York and London, 1934.

Barnes and Dumond, throw a bright light upon the origins and character of the abolition crusade. Here we see the movement as the outgrowth of the humanitarian-religious revivals of the early nineteenth century, spreading far wider and becoming far more influential than has been supposed. Formerly we were told that only a small proportion of the Northern people were abolitionists. These studies show that most of the members of the powerful evangelical churches became committed to the program. Some of their leaders, at least, looked forward with pious exaltation to the prospect of civil war with "the stealers of men." They laid the foundations for the organization of the Republican party in 1854 and they provided the votes which elected Lincoln in 1860. The fears of the Southerners in that fateful year seem to have been less unreasonable than we have been taught to believe.

Thus, the monographic attack upon the older history has forced reversal or revision of judgment upon almost every important point. The resulting damage to the traditional interpretation of the break between the sections is even greater than has been indicated, for it must be obvious to every member of this Association that many other very significant studies have not even been mentioned. Time has not permitted their inclusion. But there are two other interesting contributions to the subject which cannot be passed over.

In their brilliant work, *The Rise of American Civilization,* first published in 1927, Charles and Mary Beard have questioned the commonly accepted belief that the institution of slavery was chiefly responsible for the clash between the sections. In their view the war resulted from a desperate rivalry for control of the powers of the general government between the Southern planting "aristocracy," committed to a colonial economy, and the rising capitalist-industrial interest which had originated in New England and the Middle Atlantic states and was spreading rapidly during the 1850's into the Old Northwest. At bottom it was the old conflict between the principles of the Hamilton-Webster and the Jefferson-Jackson schools, and it was concerned with the demands of business enterprise for protective tariffs, a national bank for the regulation of currency, and Federal subsidies for shipping interests. This program was repeatedly defeated by the planters through their control of the Democratic party until the Southerners became embroiled with the Northern laborers and Northwestern farmers whose demands for free homesteads on the Western public lands they also opposed. When the Kansas-Nebraska Act opened the new territories to slaveowners the free farmers and mechanics, greatly alarmed lest they should be excluded by slave competition, formed a new political organization, basing it upon Jeffersonian principles and calling it Republican. These original Republicans in 1856 demanded only that slavery be kept out of the territories; they would leave it alone in the states. This quarrel, therefore, was over the Western lands; it was fundamentally economic and the right or wrong of slavery was not the basic issue. Moreover, the Beards say, the abolitionists were too weak numerically to have much political influence.[160] The original Republicans were not strong enough to win in

---

[160] It must be remembered that when this was written the work of Gilbert H. Barnes had not appeared.

1856, were in danger of dissolution by 1859, and were able to win the election of 1860 only after they had effected a combination with the old Whigs of the East on a platform which joined the business program with that for free homesteads in the West. The Southern planters, finding their economic order threatened by their loss of political power, resolved upon secession; but they appealed to the fellow Southerners to support them on the ground that the constitutional rights of the states and the safety of their local institutions were in danger. The Beards claim that there was no danger since Lincoln received only about forty per cent of the total votes and his party platform contained no threat of attack upon slavery within the states. With more recent evidence before us, it seems clear that these authors underestimated the strength and the strategic position of the antislavery forces; and it would be easy to raise objection to some of their statements on other points. But there can hardly be any doubt that they have made an extremely valuable contribution to the solution of this vexing problem by their emphasis upon the economic rivalries which to a very great extent motivated sectional antagonisms from 1820 to 1861.

In a recent article[161] Avery O. Craven has brilliantly presented the thesis that the clash of the sixties was the result of emotional appeals carefully nurtured, joined with sublimated economic motives, and cleverly developed into intersectional hatred. From the side of the North it came, he thinks, from the fusion of three ideas:

---

[161] "Coming of the War Between the States, An Interpretation," in *Journal of Southern History* (Baton Rouge, 1935-), II (1936), 303-22.

(1) the religious-humanitarian movement which began early in the century and gradually centered upon slavery as the most grievous of sins; (2) the conviction which grew up among the hard-pressed farmers and mechanics that the realization of democratic ideals was being thwarted by a selfish aristocracy, which came to be identified with Southern planters living in luxury off the labors of hapless slaves; (3) the belief that the economic progress of the North and Northwest—protective tariffs for the benefit of labor, internal improvements for both farmers and tradesmen, and free homesteads in the West for farmers and mechanics—was being checked by these same aristocratic and sinful planters. God's plan for an ideal democratic America, in which His elect were to be sure of profits and free homesteads, was being thwarted by these Southern agents of the Devil whom it was a Christian duty to hate and over-throw. In the South, on the other hand, the leaders, relying upon a strict construction of the Constitution to protect their staple-producing agriculture against the Northern demands for tariffs and national banks, had been unable to arouse any emotional response from the masses of farmers. While the rise of the slavery issue had made the nonslaveholders as well as the planters uneasy, because it threatened racial and social disturbance, it was not until the John Brown raid and the startling evidence of wide Northern approval of his enterprise had convinced them that they were on the brink of racial war and social chaos that the masses of the Southern people began on their own account to conjure up devils in the form of abolitionist Black Republicans. Then they were ready to fight if necessary to preserve their homes against this threatened danger. Men

on both sides had "associated their rivals with disliked and dishonorable symbols and crowned their own interests with moral sanctions."

Whether or not we have approximated to something like the final verdict on the general causes of this greatest tragedy of the American people, it is evident that we have come a long way from the explanations of such postwar historians as von Holst and Schouler. Perhaps one day the textbook writers will perceive the change. If one may dare to forecast, it seems likely that, allowing for individual variations, historians will come to agree that the break between North and South came from emotional disturbance over moral convictions (without reference to the essential quality of those convictions) and from economic rivalries, while politicians, intent only upon the immediate objective, fomented for their own ends the emotional forces which they could arouse but could not check when the crisis came. Slavery was a primary cause, but not in the sense that the older historians have made familiar. It was, as the Beards well say, no simple, isolated phenomenon.

Thus far we have considered only the newer interpretations of the forces which brought the two sections face to face in the late winter of 1860-1861 with rival governments in Washington and Montgomery. There is still much investigation to be done on the question of how the war actually came about, a matter which is likely to undergo as radical a revision as that to which the antebellum period has been subjected. For the present, however, this subject may be dismissed with the observation that the popular fear of war which had accompanied the secession of the cotton states in

December and January had subsided to a marked degree by the last of March, 1861, and that the mass of the people, North and South, then seemed confident that hostilities would be avoided, in spite of the activities of a war faction in each section; and, further, that if secession were inevitable—there are strong reasons for so thinking—the war itself was not inevitable, unless those whose official positions gave them the power to choose war or peace were obliged to choose war for reasons of policy.[162] It is sufficient here to call attention to the difference between the causes of the first crisis, which resulted in the secession of the cotton states, and the handling of the second crisis which immediately precipitated the war.

Since the purpose of this paper is only to trace the changes in the interpretation of the causes of the Civil War, its consequences and its place in American history, nothing will be said of the progress or conduct of the struggle itself. We may now consider what it has meant to the people of the United States.

When Northern people began to consider the fruits of

---

[162] In a paper, as yet unpublished, entitled "Lincoln and Fort Sumter," which was read at the meeting of the American Historical Association in Chattanooga, December 27, 1935, I reviewed the evidence which, in my opinion, shows that Lincoln himself determined upon resort to force as the only means by which to extricate his administration from a dangerous dilemma, and, in order to fix the burden of war guilt upon the Confederates, with consummate adroitness maneuvered the Confederates into "firing the first shot" by attacking Fort Sumter. He thus gained a great moral-strategic advantage in that he was enabled to appeal effectively for Northern support of the war on the plea that the "rebels" had wantonly attacked the government.

the four years of warfare, it was natural that they should think in terms of what they believed they had fought for. One large group had regarded the preservation of the Union as the main purpose, while another had insisted that the destruction of slavery was the most important aim. Before the end, President Lincoln had managed to combine both purposes. Other and minor considerations were either kept in the background or were overlooked by the majority of the people. After it was over, these two things were held to be the great achievements, worth all the cost of life and property. They had acquired in the course of the conflict an immense emotional appeal, so that a great political party, for more than twenty years after the surrender of Lee, made its appeal to Northern voters for further tenure of power upon the plea that it had saved the Union and struck the shackles from the helpless slave. It was but human nature that a people, after a successful war, should prefer to interpret it as something glorious, reflecting honor and credit upon themselves. Somewhat later, as opportunities arose for the commercial exploitation of Southern agriculture and natural resources, beneficent consequences to the South itself began to be pointed out. The "poor whites" of the Southern hinterlands had been freed as well as the slaves, and fresh currents of energy had been turned into the stagnant pool of Southern economic life. And these things are repeated unto this day. Deferring for the moment other consequences of the great intersectional conflagration, let us consider briefly these claims.

The Union was preserved, if we mean that the people of the Southern states were forced into subordination to the government at Washington. But it was twelve years after the last ragged Confederate laid down his arms before all of these people were permitted to govern themselves again, and even then the old Union had not been restored. The harshness of the victors to the conquered had aroused resentments that lasted for a generation and are traceable even now. Fortunately, however, a general reconciliation was effected in the course of time. Southern political leaders in Congress found it necessary to proclaim not only their loyalty to the nation but also that the results of the war were "all for the best." Partly out of sheer weariness of the long wrangles, partly out of a sort of fatalism, and partly because of their absorption in material interests, an increasing number of the Southern people acquiesced in the verdict. In the North, while the war feeling was periodically revived by political waving of "the bloody shirt," this appeal gradually lost its force. Prosperity and political power turned the thought of the people away from the past and made forgiveness easier. Some Southerners remained "unreconstructed"; some Northerners never gave up their suspicions of all things Southern. But the flowing years smoothed the rough edges of mutual distrust.

But there are other considerations which the historian should take into account. Some of them were examined not long ago by Professor Richard H. Shryock in a very thoughtful paper entitled "The Nationalistic Tradition of the Civil War."[163] Why was this Union so valuable that it was worth the lives of more than half a million men to preserve it, aside from the vast amount of wealth destroyed and the untold suffering left in its train?

---

[163] *South Atlantic Quarterly* (Durham, 1902), XXII (1933), 294-305.

Americans have been prone to boast of the size of their country. Is there some mystical, imponderable but precious quality in mere bigness? Or was it the danger (as argued by Lincoln) that the successful withdrawal of the Southern states would lead to the secession of other states, so that the once mighty Union would have been broken up into a group of petty, mutually jealous and possibly warring little countries? Of that there seems to have been no great danger, for the free states were already bound closely together by economic ties, as well as by blood relationships. The most that could have been expected was that two or three of the border slave states would have decided to join the cotton Confederacy, or that adjustments of boundary or customs lines might have given trouble for some years after 1861. While it is clear that the political unity of all the states has been an economic blessing to those who have found their best markets within the United States behind a high protective wall, it would be hard to prove that these benefits have been shared by all sections, or by all economic groups. As for the assertion so often made that mutual jealousy and fears would have made huge standing armies necessary, we have in refutation the example of our happy relations with Canada. Furthermore, who can say that, had no war been made on the seceded states, there would not have been quicker reconciliation and reunion on a firm and mutually satisfactory basis?

But some may say, "Would not slavery have remained to plague them all, to keep alive reproaches and bitterness, to endanger peace? And could the Confederacy have prospered with its agriculture and industries hampered by a system so outworn and wasteful?" We can all agree that not merely the Negro but the Southern white man and the South as a whole is better off with slavery gone. But this admission does not precisely meet the question: Was this the only or the best way to get rid of the institution? And was the manner of it worth the cost?

There can be little doubt that the institution of chattel slavery had reached its peak by 1860 and that within a comparatively short time it would have begun to decline and eventually have been abolished by the Southerners themselves. The extraordinary expansion of cotton acreage would almost certainly have resulted in overproduction and lower prices, and the rapid introduction of laborsaving machinery in every step of cotton production except picking would have made slaveholding too expensive for the great majority of planters. Only those who could be surest of continued profits—the sugar planters, perhaps, and the owners of the most fertile cotton lands—could have afforded the expense of rearing their full supply of labor; they would almost certainly have discovered that it would be cheaper to hire labor as needed. And the problem of controlling the freed Negro could have been worked out on the basis of the existing laws. But it is not profitable to speculate too minutely upon these probabilities. The prospect of social evolution was not given a chance.

Chattel slavery was destroyed in the roar of cannon and the murderous rattle of musketry. But the central problem, the adjustment of two intermingled but dissimilar races, was not solved. It remained to plague not only that generation but future ones, and was made far more difficult by the alienation of the races which resulted from the political and social experiments of radical reconstruction. Nor can we overlook the actual suffering from disease, malnutrition, and the other ills which befell the unfortunate blacks during the tumultuous period of transition from slavery to freedom. Beyond question the Negroes have made substantial progress since 1865, but that their condition is better now than it would have been had a more orderly process of social and economic change been followed is not so certain. No thoughtful friend of the Negro will affirm that the economic freedom of the race is yet in sight. The results but illustrate anew the truth that every radical reform in our complex society brings new problems in its train.

It is a favorite dictum of many writers, more especially the Northern historians, that the emancipation of the African also set free the "poor whites" of the South. Usually these writers seem to regard as "poor whites" all who were not slaveowners. For something like a hundred years Southerners have been trying to make clear to Northerners the falsity of this definition and the difference between "poor whites" and the great middle class of nonslaveholders. For some reason the misapprehension survives and it would be too wearisome to explain the distinction again. But is it true that the "poor whites" of the South (as Southerners use that term) have been appreciably uplifted in the economic and social scale by the Civil War? That the descendants of some of those who lived in the lower fringe of the social order in the Old South have greatly improved their condition is unquestionable; but that many of this class have furnished leaders to the New South would be extremely difficult to prove. While thousands of sons and daughters of poor families have risen to prominence in business and the professions (as others did before 1861), very few of them came of what the Southerner has always called "poor white trash." This submerged class remains submerged. We have but to look at the white sharecroppers throughout the cotton country and the white casual laborer and ask ourselves whether these people are appreciably better off than were their ancestors in 1860. It would not be difficult to show that in these submerged groups are many families descended from the independent yeomanry of 1860 who have sunk to their present status under the pressure of adverse economic conditions. The condition of the white farm tenants lends little support to the thesis that the Civil War was a boon to them.

But, it is said with the fervor of conviction, at least the breaking of the power of the planting aristocracy opened the way for industry and commerce and the economic regeneration of the region. Look at the industries which have spread along the piedmont from Virginia around through the Carolinas, Georgia, Alabama, and up through Tennessee and Kentucky. Look at Atlanta and Birmingham and the scores of other thriving cities which remained but small towns as long as Cotton was King! The obvious retort is that the destruction of war hindered rather than helped this development. The researches of the younger historians are making it clearer every year that, while Southern industry was in its infancy in 1860, the infant was a very healthy one and gave promise of rapid growth. The war destroyed most of what had been built and the capital of the builders was, in most cases, wiped out. When the new start was made railroads, mines, and factories recovered slowly and painfully except where Northern capital was available; and the price of Northern assistance was usually the surrender of control and of the larger share of the profits to residents of New York, Boston, and other Northern centers.

We are told that the defeat of the secessionists not only kept the Union intact but welded the people of the United States into a nation; and in an age when the trend to nationalism throughout the world has been so strong this achievement has seemed very important. The Federal armies destroyed the Calhoun theory of the sovereignty of the states and gave the sanction of victory to the Webster-Lincoln conception of national sovereignty. More specifically, the three "war amendments" to the Constitution enlarged the powers of the general government at the expense of the states. Of special significance has been the "due process of law" clause in the Fourteenth Amendment which, as interpreted by the courts, has hampered the states in their efforts to tax and regulate the great corporations. But while we grant all this, must we assume that civil war was necessary to achieve nationality or to bestow needed authority upon the general government? Are there not plenty of evidences that before 1860 powerful factors were already working toward greater economic and social unity, even though obscured by the sectional quarrel? The railroad, the telegraph, the press, and technological advances were extending business enterprise into wider fields. Farmers, business men, and even laborers were becoming conscious of their group interests. Shall we deny that this movement would have gone on in the same general direction that it has taken since 1865? Would not the Federal government have been called upon eventually to deal with the problems raised by these groups, expanding its powers to that end by interpretation and amendment of the Constitution, as it has done?

Considerations such as these have led some scholars, chiefly of the Turner school, to doubt whether, after all, the Civil War changed the course of American history as much as we have been accustomed to think. They point out that the fundamental forces shaping American life were neither changed nor greatly affected by the war. The frontier line continued to roll westward, railroads were already tending to consolidation and cheaper transportation, and to plot their courses toward the Pacific. The colonial economy of the Southern planters was destined to decay in any case, the slaves must have been emancipated in the course of time. The armies of industry and labor were gathering their forces in textile, steel, coal, oil, and a hundred other camps before the raw troops of McDowell and Beauregard met on the Plains of Manassas; the rise of the mechanical industries to a dominant place in our national economy was as inevitable here as it was in Western Europe. On the other hand, Charles and Mary Beard are of the opinion that only the destruction of the political power of the obstructive planter class gave room for the rising new business enterprise which, directed and controlled by the ruthless captains of industry, proceeded to change the whole structure of our economic, social, and political institutions. But the two views are not irreconcilable. Is it not possible to agree that the swing to industrialism was inevitable and to agree also with the Beards that by suddenly sweeping away the impediments to business enterprise, and by preventing for forty years effective control of individualism-run-mad, the war helped to create a host of intricate problems which we have not yet been able to solve? In short, did not the Civil War, indirectly perhaps and at long range, help to get our own generation into its perplexing predicament?

Now that we look back with the advantage of more than seventy years' perspective to the great conflict in which our fathers and grand-fathers fought and seek to analyze the causes and appraise the consequences of the desolating struggle, what conclusions seem finally tenable? The forces that swept toward the disruption of the old Union were far more complex than contemporary observers or the early historians perceived. The slavery issue remains a prime factor, but not in the simple terms set down by von Holst, Schouler, and Rhodes. It was in itself a very complicated issue and it was interwoven with a mass of other complexities. Clashing economic interests, and, to an extent, political ambitions played their part. And not merely stubborn differences of opinion on trivial or fundamental policies made adjustment difficult, but likewise the emotions aroused by mutual misunderstanding and the fear of responsible political leaders lest they, by concessions, lose the confidence of those whose emotional support they had enlisted. The statement that the war was "to save the Union" ignores all the forces which had brought the two sections into hostility. The phrase "to destroy slavery" is either a confession or an afterthought. And what have been the consequences which, after all, must give the conflict its place in the story of American development?

Making all necessary allowances for our inability to weigh accurately the imponderables in the history of a great people, can we say with conviction that this war accomplished anything of lasting good that could not and would not have been won by the peaceful processes of social evolution? Is there not ground for the tragic conclusion that it accomplished little which was not otherwise attainable? Had the more than half a million lives and the ten billions of wasted property been saved, the wealth of the United States and the welfare of the people would not likely have been less than they are now. Perhaps some of the social and economic ills that have bedevilled us for the past fifty years would have been less troublesome.

Will such reflections enable us to attack our present problems with less of emotion and more of cool reason than we frequently display? That, at any rate *should be* one of the lessons of History.

Some Problems Involved in Writing the History of the Confederacy[164]

by

Charles W. Ramsdell

> *We greatly need more good monographic studies based upon an exhaustive examination of sources. The very complexity of the field as a whole presents a difficult problem in the organization of the material. The evidence on many points is very scanty and in some cases is likely to remain so; in other instances, though fairly abundant, it is often technical or conflicting. After all, however, these problems are always present to worry the historical investigator when he attempts to cover any large field of human endeavor.*

When, unguardedly, I succumbed to the blandishments of the chairman of the program committee, Professor Binkley, and agreed to present this paper, it seemed that

---

[164] Charles W. Ramsdell, "Some Problems Involved in Writing the History of the Confederacy." *The Journal of Southern History*, Vol. 2, No. 2 (May, 1936), 133-147. Article and citation are verbatim.

the task would be a fairly easy one.[165] But later, when I attempted to set down explicitly these "problems" as I see them, it became painfully evident that they are only my problems and that they may not present the same difficulties to others that they do to me. While, therefore, this presentation may prove to be only an embarrassing confession of my own ignorance, perhaps it may still serve its purpose by provoking a general discussion which will elicit new points of view and new sources of information. Of course, everyone here will understand that it is impossible within the limits of such a paper as this to list all the problems which must confront anyone who undertakes such a complex subject as the history of the Southern Confederacy.

This future historian will first have to answer for himself the question, "What kind of a history is this to be? What is to be its scope, where its emphasis?" If we answer the question for him, we shall probably say that what we do *not* want is a history that lays undue emphasis upon any particular phase of the story, whether it be military operations, or the political and administrative policies and difficulties of the Confederate government, or the socio-economic conditions of the people. We want instead a full, comprehensive, well-balanced and articulated account that will give due weight to all discoverable factors in the struggle of the Southern people for independence and their failure to achieve it. Without sacrificing accuracy, it should have as much literary charm as the writer is capable of imparting to it. You may

---

[165] This paper was read before the Southern Historical Association at its First Annual Meeting, in Birmingham, Alabama, October 25, 1935.

say that such an ideal is unattainable, and it probably is; but our historian must necessarily set up some such ideal even though he may not be able to approximate it. It becomes evident at once that he will need our sympathy.

We may assume that our historian has already done extensive research in the period of the Confederacy. If he is at all well qualified for his task, he will have discovered that there is hardly a problem of any consequence whose roots do not run back into antebellum conditions. He must therefore make himself as well acquainted as possible with the antebellum South. It need not be said to this audience that, although much excellent work has now been done in that field and more is under way, much more still needs to be done before we can visualize the whole picture of the South as it was just before secession. Thanks to the great work of the lamented U. B. Phillips and others we now have illuminating accounts of the organization and administration of the plantation and the working of the system of slavery. This will be helpful in dealing with one important aspect of the life of the Confederacy. But we know too little about the outlook and attitudes of the small farmers who constituted the great mass of the population. It would be a great boon to have such a study of the antebellum small farmer as Rupert B. Vance has made of his descendants; but the material for case-histories is lacking. Too little is known about the Southern business men, whether merchants, factors, industrialists or bankers-men who were to play important, if relatively inconspicuous parts in the struggle for Southern independence. Broadus Mitchell, in his *William Gregg,*[166] and Dr. Kathleen Bruce, in her

[166] Chapel Hill, 1928.

excellent *Virginia Iron Manufacture in the Slave Era,*[167] have shown what may be done on the little known subject of Southern industrial development. But our historian will need to know something of the status of the iron business in northern Georgia, Alabama, and Tennessee, for it was from the little mines and smelters in these states that the Confederate ordnance bureau was procuring the greater part of its iron by 1863. And it would be helpful for him to get as much information as possible about the little wool and cotton yarn mills and cloth mills which were scattered from Virginia to Mississippi, inadequate though they were for the needs of both the army and the civilian population. Especially useful would be some knowledge of Southern banks and banking, their functional relation to both Southern and Northern credit policies, and of the banking laws of the several Southern states and how they actually operated. He may then be able better to understand the wartime financial and fiscal policies of the Confederate government and of the states. In brief, he must acquaint himself with the nature and extent of the material resources of the South before he can proceed to the more difficult task of discovering and revealing how they were organized, administered, and utilized under the Confederacy. And at present he would find himself sorely in need of searching studies of antebellum state politics—studies which give careful attention both to personal and factional rivalries and to economic and social backgrounds. For some states virtually nothing of lasting importance has been done while for others such studies as have appeared have tended to present the reaction to Federal politics rather than to local issues.

---

[167] New York, 1931.

This is not to intimate that interest in Federal issues was unimportant, for it *was* important; but one suspects that local or intrastate issues played a much larger part in state affairs even in the eighteen-fifties than they do in most of our state histories. And we may be quite sure that local antebellum political alignments and jealousies were carried over into the public affairs of the Confederacy. All these things are mentioned merely as illustrations of the general contention that our unfortunate historian will need to know much more than any one of us, I suspect, now knows about the antebellum South.

Coming now directly to the Confederacy itself, let us consider for a moment the problem of handling the military operations. Notwithstanding the tendency of historians in recent years to relegate military matters to a much less conspicuous place in favor of economic, social, and other factors—a shift of emphasis with which, I confess, I strongly sympathize—how can any comprehensive history of the Confederacy neglect the military portions of the story? Popular attention was fixed upon the armies as upon no other one thing. Upon them depended the fate of the "revolution." To use a very trite illustration, we cannot eliminate the melancholy Dane and still call our play "Hamlet." But when our historian undertakes the story of the military campaigns he must face two difficult problems. One is that of space and proportion. It is extremely difficult to describe such a complicated thing as a military campaign briefly and also satisfactorily, for almost innumerable factors arose to condition every plan and movement. If the historian says nothing of them he leaves the impression that the armies moved in straight lines over a smooth surface, and he really tells us nothing. If, on the other hand, he tries to tell everything that had a significant bearing on the results of the operations, his narrative runs on and on into more volumes than he can bear to think of. The other difficulty is likely to be even greater. It is that of so analyzing the military operations that he can retain the respect of competent military technicians. It may seem surprising that out of all the extensive literature of the Civil War very little has been written in a manner to satisfy the critical expert who is thoroughly trained in the techniques of both military science and historical investigation. Very few of the narratives of participants can be relied upon; for some had forgotten much before they began to write; others wrote primarily to defend their own reputations; others, still, merely to meet public demand or to gratify a very human desire to leave a record of their own achievements. When they consulted documents they used official reports, a notoriously faulty type of evidence. Later historians have too often relied upon these same narratives or official reports. Most of them lacked the technical training for analyzing a military situation or the logistics and tactics employed and have given no consideration, or very little, to conditions of terrain, weather, roads, means of transportation, availability of supplies, and the scores of other things that contributed to success or failure. There are some men, trained in the staff schools or the War College of the United States Army, competent in both the military and the historical techniques, like Colonel A. L. Conger and Colonel O. L. Spaulding, Jr., who have made permanent contributions to the study of these military operations; and there are others in civil life like Thomas R. Hay and

Douglas S. Freeman whose fine work is fresh in our minds. There are some whose technical knowledge of military matters cannot be questioned but who have failed utterly to take into account the imponderables that weigh so heavily in warfare and, which is even worse, have shown a deplorable lack of critical ability in the handling of evidence. The work of the distinguished British officer, Major General J. F. C. Fuller, may be cited as an example.

What is our historian to do if he lacks the requisite training in military science? It is easier to ask the question than to answer it. Perhaps his only safety lies in first getting the critical advice of experts and then in resort to caution and prayer.

There is another subject closely related to military operations which has received little attention from the military historians; and this is surprising because every trained military man is well aware of its importance. I refer to the work of the services of supply or, to use the terminology of the 1860's, the subsistence, quartermaster's, medical, and ordnance bureaus of the war department. In the first place, these bureaus were absolutely essential to the very existence of the armies and any serious lapse in their functioning might quickly disrupt the plans of the military commander and involve the loss of a campaign. When the bureau of subsistence failed to provide food for the men, or the quartermasters to furnish shoes or sufficient transportation, or the ordnance bureau to bring up ammunition, even military genius could not overcome such a handicap. And such things happened. In the second place, these services of supply reached into almost every community of the South and out through the blockade to Europe. The inside story of their administration, if it could be fully told, would not only throw new light upon the difficulties and some of the failures of the commanders in the field but would also reveal much about the resources of the Southern people and the troubles encountered in making them available. In fact, they touch upon almost every activity of the Confederate government and on much of that of the states. But the difficulties of reconstructing even an approximation of the full story of these services seem practically insuperable. That vast compilation, *The Official Records of the Union and Confederate Armies,*[168] though an unrivaled storehouse of information, by no means contains adequate material on this subject. Compiled by war department clerks who had no training in and little conception of the importance of the economic side of war, it places the emphasis upon the work of the armies in the field. Such information as pertains to the Confederate services of supply seems to have been included rather incidentally. This is especially true of the bureau of subsistence which was presided over until near the end of the war by the eccentric L. B. Northrop. Many of the original records were lost during the conflict or were destroyed, accidentally or purposely, in the confusion of the last days. Some of them, overlooked by the compilers of the *Official Records,* are to be found in the "Confederate Archives," Old Records Division of the adjutant general's office in Washington; but they are difficult to search out. A little material has found its way into other public repositories, a little is in private hands. But unless other collections are uncovered, there will

[168] Washington, 1880-1901, 70 vols. in 128.

remain many puzzling gaps in the story.

Everybody knows that one of the heaviest handicaps of the Confederates was their lack of sufficient mechanical industries to supply their own needs. Reference has already been made to the desirability of learning more about the extent and conditions of these industries before the war began. It is, of course, even more necessary for the historian to find out all he can about their condition and contributions to the common cause during the war. Of some we can find incomplete accounts, amounting in certain cases to mere fragments; of others merely the location; while others still have left little more than a trace. Some, we know, wore out their machinery and closed down; some were destroyed, with all their records, either by accidental fires or by invading armies.

We can only hope that sometime enough records may turn up to enable the historian to reconstruct their story in greater part than now seems possible. We know that there was a deplorable scarcity of every kind of fabricated article—cloth, tanned leather, iron or steel tools, horseshoes, plows, nails, needles, bagging and rope for cotton bales, sacks for shipping grain, glassware, everything that was in common use. One of the prime necessities, salt, now so common that we take its abundance as a matter of course, was so scarce that procuring even a meager supply became one of the major problems of both the people and the state governments. Fortunately we now have the excellent study of Dr. Ella Lonn on *Salt as a Factor in the Confederacy.*[169] But we lack any sort of account of such local industries as tanneries, wagon shops, shoe shops and the like. The

---

[169] New York, 1933.

same is true of the small private gun factories; but thanks to the enthusiastic researches of a few collectors of old firearms the locations of many of these small factories have been determined. Too often, however, the quantity and quality of their output have remained mysteries. It may seem to some of you that I am asking too much of the historian of the Confederacy in setting this problem before him. I am convinced, however, that the scarcity of fabricated articles so much needed in everyday life, had an important bearing on the outcome of the war, for it affected not merely the efficiency of the armed forces but had much to do with the war-weariness and the irritation that were so much in evidence in the last two years of the conflict. Our historian should neglect nothing that materially affected the conditions and the temper of the people who must support the armies.

The development of the railroads in the antebellum South has received some attention, beginning with U. B. Phillips' *History of Transportation in the Eastern Cotton Belt* in 1908;[170] but except for one brief article published nineteen years ago and the interesting but rather superficial chapter in F. B. C. Bradlee's *Blockade Running during the Civil War,*[171] there has been very little published on the part played by the railroads in the struggles of the Confederacy. Mr. Robert S. Henry's *Story of the Confederacy*[172] is practically the only narrative of the military campaigns that gives attention to the problem of Confederate railroad transportation. There is considerable material available, but it is scattered through

---

[170] New York, 1908.
[171] Salem, Mass., 1925.
[172] Indianapolis, 1931.

almost every conceivable collection of sources. The material on inland waterways transportation—by river and canal—is more restricted and more difficult of access. Of wagon roads and wagons our historian will probably be able to say only that the roads were usually very bad and that wagons, and teams to pull them, became so scarce that local transportation in many sections broke down completely.

Of paramount importance in any general account of the Confederacy are the financial and fiscal operations and devices of the general and the state governments. All of us are familiar with the work of J. C. Schwab, *The Confederate States of America,*[173] and of E. A. Smith, *The History of the Confederate Treasury,*[174] both of which appeared in 1901. These men actually examined the records of the Confederate department of the treasury and their publications are very helpful in following out the formal operations of that department; but they wrote at a time when economists were very conservative and they were both much concerned with pointing out the dangers of paper money inflation. They did not, I think, give sufficient attention either to the background of Southern conditions—I mean the credit system, the tax systems, banking practices and banking laws before secession—or to the inherent difficulties which confronted the treasury officials during the war itself. Anyone who looks into all the pertinent facts without prejudice will have some difficulty, I apprehend, in determining just how the Confederacy could have solved its financial problem. Some errors of the Confederate government, very serious

---

[173] New York, 1901.

[174] Harrisburg, 1901.

ones, are obvious. Certainly the refunding act of February 17, 1864, proved worse than a failure; but was there really any way by which at that late day the Congress could have redeemed the credit of the government? In this connection, it would be well for the historian to look into the banking experience and the financial views of Christopher G. Memminger before the war as he expressed them in his controversy with the Bank of the State of South Carolina. If he thinks that Mr. Memminger was entirely ignorant of the principles of public finance he may find things that will surprise him. The development of the Confederate tax system as well as the tax systems and financial operations of the several states and their repercussions upon Confederate finances will require careful consideration.

Among the many measures of the Confederate government which must receive attention are the following: the provisions and administration of the conscription acts; the methods adopted to check desertion from the armies (and the causes of desertion); the suspensions of the privilege of the writ of habeas corpus; the sequestration of the property of "alien enemies"; the policy and methods of impressment of supplies and labor; the belated efforts to control blockade running in the interest of the general government; and the effects of all these measures upon the public temper, or, more accurately, upon the interests and attitudes of certain economic and political groups of the population. Fortunately, there are excellent monographs on some of these subjects: Dean A. B. Moore's *Conscription and Confiict;*[175] Dr. Ella Lonn's *Desertion during the Civil*

---

[175] New York, 1924.

War;[176] and Mr. S. B. Thompson's *Confederate Purchasing Operations Abroad.*[177] Although the historian may be able to bring to light some new material on the subjects of these special studies, he is not very likely to find any that will warrant a substantial modification of the conclusions set forth in them. Evidence is available on all the others, although scattered sometimes in rather obscure places. The task of our historian will be not so much to find the sources of information as to weigh the relative effects of all such measures upon the work of the armies, the condition and attitudes of the people and the fortunes of the Confederacy. In doing this he will find difficulty enough to challenge all his powers of analysis.

There were two other difficult problems of the Confederate government—and, in large measure, of the states also—which are likely to prove nearly as troublesome to the historian. One was how to make best use of the slave population; the other was what to do with the great staple crops, especially cotton and tobacco. As to the slaves, we know that it was expected at first that they would be kept at their usual tasks on the plantations or at other work, except when hired as laborers on fortifications or in some other capacity with the armies. Even free Negroes were not to be accepted as soldiers. But slaves on the plantations must be kept under the accustomed discipline if they were to produce the needed food supplies. There was little difficulty on this score until the Federal armies made lodgment on the coasts and began to penetrate the planting regions of the Mississippi Valley. Then the Negroes became restless and the

---

[176] New York, 1928.
[177] Chapel Hill, 1935.

plantation routine near the Federal lines began to break up. This situation was the chief cause for the passage of the famous "twenty negro" exemption law of October 11, 1862, which had such troublesome repercussions among the nonslaveowners that, after some modifications, it was replaced, in February, 1864, by the provisions for "bond-exempts." But this was no less unpopular among the poorer people. To what extent did the poorer men who raised the cry, "It's a rich man's war and a poor man's fight," merely seize upon this particular exemption as a method of justifying, or of rationalizing, their very natural desire to escape the dangers and hardships of military service? Was the law unjust and inexpedient, considering the fact that food supplies must come chiefly from the plantations which could produce a surplus instead of from the small farms which could furnish but little? How else could the government, in the light of what we know of plantation management, have solved the problem? One other aspect of this slave problem should be mentioned. As the war dragged on and replacements in the army became more and more scanty, the government endeavored to make wider use of Negro slave labor on fortifications. Failing to procure enough by hiring, it resorted to impressment; and several of the states also passed laws for the impressment of slaves. Most planters protested the taking of their Negro hands, alleging that the slaves were often overworked, underfed and were generally badly treated by the army officers, and that the withdrawal of their laborers from the plantation at critical times in planting, cultivating and harvesting crops was ruinous to the production of food supplies and therefore to the cause itself . How much of truth and how much of rationalization was there in all this? Or was the trouble merely in the administration of the law?

The cotton and tobacco question presents many angles. Could the Confederate government have utilized all the supply of cotton as a basis of credit in Europe at the beginning of the war and thus have solved its financial and naval problems? I mention this question not because I think it debatable—for I am unable to see that the scheme was feasible—but because it has been raised by others from time to time and it offers a neat opportunity for a discussion of all the factors involved in the situation. Let us turn to another side of the cotton question. Very early in the war the government undertook to prevent its exportation through the military lines to the North. For a time such cotton as was in danger of capture was burned either by those owners who were sufficiently patriotic or by state or Confederate officers. When the patriotism of the owners weakened because of their increasing privations and eager Northern buyers appeared just beyond the enemy's lines, a furtive but brisk trade through the lines began. The United States government encouraged this trade while the Confederate authorities frowned upon it and endeavored to stop it; but it went on throughout the rest of the war wherever the Union armies reached the cotton country. It made enormous profits to traders while to many Southern families it was the only means of procuring the necessities of life. The situation was one to foster official corruption; but although the air was full of accusations and innuendoes the beneficiaries were generally powerful enough to cover their tracks and but little direct evidence implicating individuals has come to light. Perhaps it should be said at this point that most of the evidence of official corruption points to officers in the Federal rather than those in the Confederate army; but that may have been because of the difference in opportunities. It would be interesting to know even approximately what amount of supplies for the Confederate armies came through the lines from New Orleans in return for cotton and sugar while Benjamin F. Butler and Nathaniel P. Banks were in command in the Crescent City. Such exchanges went on in other quarters also but the evidence is generally so fragmentary that it excites without satisfying one's curiosity. One would like to know also what effect this clandestine trade between individual Southern families and Northern agents had upon the loyalty of those Southerners to the Confederate cause.

The limitations imposed by individual states upon the planting of cotton and tobacco, and the prohibitions upon the distillation of liquors, were interesting experiments in a region where the private property rights of the individual had always been held sacred. The laws were based upon the public necessity for the greater production and conservation of food supplies. It is easy to find what these laws provided; but it is very difficult to determine how well they were obeyed and administered and how effective they were.

The subject of the clandestine cotton trade suggests another. What do we know of the condition of Southern families left within the Federal lines in the subjugated districts, especially along the Mississippi? How did they adjust themselves to this situation? What of the treatment of the Negro population in such districts by Federal officials and by the favored contractors who were given the privilege of working abandoned plantations? Or should the historian of the Confederacy confine his attention to the ever-narrowing territory within the Confederate lines?

If he attempts to describe business conditions in the South during the war, what can he say except that most of it was deflected from its usual channels, that along some lines it practically dried up while in others it was stimulated to an extraordinary degree? We know in general the effects of the blockade, the downward plunges of the currency, the breakdown of transportation facilities, the frequent impressments, occasional state embargoes, the irruptions of Federal armies and the devastations of large areas in northern Virginia and in the Valley of the Mississippi. But very few records of business firms during this troubled period are available and most of these are fragmentary. No business statistics seem to have been gathered in sufficient quantity to be of much value. In short, we are forced to resort to general deductions or impressions; for while we feel reasonably certain about some of these things, we lack the detailed evidence with which to support our generalizations. We find that some firms, fortunately situated, made enormous paper profits. We find much complaint of "speculators" who forestalled the markets, monopolized the necessities of life and callously oppressed the poor; but it is not always easy for us to distinguish between what was indubitably profiteering and what was the inevitable result of the rapid fall of the currency or of the actual physical scarcity of goods. Blockade-running made huge profits and the stocks of the corporations engaged in the business were bought and sold with all the frenzy that ever characterized the New York Stock Exchange; but extremely little information about the financial operations of these companies seems to have been preserved.

We know, likewise, that the families of the poor often endured the most severe privations, whether they were soldiers' wives and children living on small farms out in the remote hills or town dwellers trying to eke out an existence on fixed incomes whose purchasing power had vanished. We have some records of the townsmen's difficulties, but the rural folk were not the sort to leave much in the way of records. They must have written to their menfolk in the armies, but soldiers in the field could not preserve letters for the future historian.

This paper is becoming too long, but something should be said about the matter of politics. Despite the assertion of some Confederate leaders that politics was adjourned during the struggle for independence, one does not have to go far into the records to discover that this was not the case. As a matter of fact—and it would have been strange if it had been otherwise—when the first Congress met in Montgomery in February, 1861, it was divided into mutually distrustful groups. After the policies of the Davis administration had begun to take form, there was a marked shifting of old party lines, with former Whigs and Democrats supporting the administration and other Whigs and Democrats opposing it. There are, here and there, very definite traces of old divisions and old animosities, but the tendencies were for new alignments. In many instances these new groupings are very hard to trace, so many are the crosslines of obscure individual or local interests and so scanty is the documentary evidence.

While we can follow the votes as recorded in the *Journals of the Confederate Congress,*[178] the failure of the Congress to record and publish its debates reduces us largely to conjecture. The Southern Historical Society has done what it could to fill the gaps by publishing such of the proceedings, including summaries of speeches, as appeared in the Richmond newspapers; but its record begins only with the first session of the "Permanent" Congress on February 18, 1862, and cannot, naturally, cover the numerous secret sessions. It seems strange that so little has been discovered of the correspondence of the members of the Confederate Congress. If a few collections of such correspondence or a few good diaries of members of the House and Senate could be unearthed, we might be able to get at the explanation of a number of puzzling things. How, in the face of the military disasters in the West and on the coast, in the face of the growing unpopularity of members of his cabinet, such as Judah P. Benjamin, Memminger and Stephen R. Mallory, did Jefferson Davis manage for so long to maintain his hold over a majority of Congress? What of the activities of the congressional cliques that centered around P. G. T. Beauregard and Joseph E. Johnston? What are the real explanations of the widespread reaction against the administration as revealed in the congressional elections of 1863? Or was it really directed against the President? There is considerable evidence of the activities of the Georgia group of malcontents, but less is known of those in other states. Every state had its antiadministration group, large or small, but in most cases the records of their plans and activities are tantalizingly scanty. How

[178] Washington, 1904-1905, 7 vols.

much of this local opposition was based upon a doctrinaire devotion to the traditional rights of states—so strongly emphasized by Professor Owsley in his well-known book, *State Rights in the Confederacy*[179]—how much grew out of personal pique or jealousy, how much came of the politician's tendency to capitalize popular discontent, how much was due to honest difference of opinion? It is a nice problem.

It is obvious to anyone who has made any study of the Confederacy that these are by no means all the special problems that will confront the historian who would tell the full story. For instance, nothing has been said of the problem of the churches; nothing of the army hospitals, nor of the search for indigenous substitutes for medicines, nor of relief organizations. I have barely hinted at some of the many social activities of the states. I have not even suggested that there were troublesome constitutional problems; nor have I mentioned the difficulty of finding the records tracing the activities of the Confederate district courts.[180] I have said nothing whatever about the foreign policy or foreign relations. This last subject, however, has received much more exhaustive and more adequate treatment than has the internal history.[181] This rambling paper must end. By way of summary, it may be said that anyone who, in the present state of our knowledge and available sources, attempts now to write a comprehensive history of the very

---

179 Chicago, 1925.

180 Since this paper was written Major William Robinson has announced his discovery of the records of these courts. He has in preparation a work on the Confederate judiciary.

181 The latest and fullest discussion is in Frank L. Owsley, *King Cotton Diplomacy* (Chicago, 1931).

complex life of the Confederacy must do a great deal of pioneer work for himself. We greatly need more good monographic studies based upon an exhaustive examination of sources. The very complexity of the field as a whole presents a difficult problem in the organization of the material. The evidence on many points is very scanty and in some cases is likely to remain so; in other instances, though fairly abundant, it is often technical or conflicting. After all, however, these problems are always present to worry the historical investigator when he attempts to cover any large field of human endeavor.

Carl Sandburg's Lincoln[182]

by

Charles W. Ramsdell

One would expect of Carl Sandburg, poet-journalist of distinction, writing of a high literary quality. Brilliant passages there are, and coruscating flashes, but the quality is uneven. Much of the writing is mere quotation, direct or indirect, sometimes thrown together rather loosely, sometimes cemented together by the author's comments. There is no such sifting, analysis, and conclusion, expressed in the writer's own language, as one finds in Freeman's R. E. Lee.

After seventy-five years the flood of writings about Abraham Lincoln which began immediately after his assassination has shown no sign of abatement. Its volume has exceeded that of publications about any other American except, possibly, George Washington. Its range has comprehended at one extreme folk-myths and savior-myths, legends, deliberate inventions, mystical interpretations, and uncritical eulogies, at the other

[182] Charles W. Ramsdell, "Carl Sandburg's Lincoln." *The Southern Review*, VI (1940-1941), 439-53. Article and citation are verbatim.

careful factual studies and a very few of the debunking type. This perennial interest in the man has resulted in literally thousands of books, pamphlets, articles, poems, plays, and stories. A few years ago, one of our foremost authorities on the Civil War, Professor James G. Randall of the University of Illinois, in a learned and witty paper entitled "Has the Lincoln Theme Been Exhausted?" surveyed the writings on the subject and the materials still to be examined and then came to the very sane conclusion that it is still far from exhaustion. Since then the presses have been as busy as usual with additional Lincolniana of varying quality. Among the most interesting items are *Lincoln and the Civil War in the Diaries and Letters of John Hay,* edited by Tyler Dennett, and *The Hidden Lincoln,* compiled from the papers of William Herndon by the late Emanuel Hertz. Now we have, in four thick volumes, *Abraham Lincoln: The War Years*[183] by Carl Sandburg, the poet-journalist, who thus completes the most extensive study of the great man yet produced. He finishes a task begun some twenty years ago, and with his two previous volumes, *The Prairie Years,* published in 1926, has given us the longest and most detailed life of Lincoln.

While the work lacks some of the *indicia* of formal historical writing, as practiced by the professional historian, it cannot be dismissed as bad history. For though one does not usually expect a critical biography from the hand of a poet who writes of his life-long hero, or an adequate understanding of all the great problems and developments which impinged upon his subject's

---

[183] *Abraham Lincoln: The War Years,* by Carl Sandburg. New York: Harcourt, Brace & Company. 4 vols. $20.

career, the reader of these volumes soon discovers that Mr. Sandburg has tried to do an extremely difficult job with entire honesty. His warm imagination, his gift for vivid and colorful description—so welcome and so rare among the "professionals"—have not been allowed to run away with the evidence. But it would be going too far to assert that he has fully divested himself of the Lincoln legend. He grew up in the midst of it; he has breathed its atmosphere all his life; and naturally he has not escaped entirely from its powerfully pervasive influence.

It may seem singular that, although several of them have written very good brief biographies of Lincoln, no professional historian has yet produced a full-length account of that very remarkable man. The late Albert Beveridge undertook the task, but death stopped him when he had brought his story down to the summer of 1858. There have been many reasons why competent scholars have hesitated to attempt such a work. For one thing, few of them have the private means which would enable them to devote the long years of labor requisite to a task of such magnitude. Then, they well know that the complete story cannot be told until the mass of Lincoln papers now under seal in the Library of Congress shall be opened to use in the summer of 1947. These papers, long in the possession of Robert Todd Lincoln, son of the President, were used by Nicolay and Hay in the preparation of their ten-volume work, *Abraham Lincoln, A History,* published in 1890, but have been inaccessible to all others. Nicolay and Hay also compiled from these papers two editions of Lincoln's writings, but omitted the much larger mass of letters and other papers addressed to Lincoln. Obviously the omitted portions are very important for the determination of influences and motives. Meanwhile several hundred other letters, papers, and telegrams of Lincoln have come to light and have been published, and others are constantly being unearthed. Another cause for hesitation is the fact that many of our former ideas about that period are being drastically revised or challenged by fresh monographic studies, some of which have to do directly with Lincoln's career. While it would be foolish to insist that the biographer should wait until all the controversial questions shall be settled—since they will never be settled to the satisfaction of everybody—the scholar who is keenly aware of them may be pardoned for his reluctance to commit himself prematurely. Still another probable deterrent has been the power of the "Lincoln legend," whose devotees will have no tampering with the thesis of their hero's perfection.

While some of Lincoln's most ardent admirers have honestly sought only the truth about him, even they have been prone to resolve all doubts in his favor on the ground that an unfavorable verdict would attribute to him motives which were "not in character." The vast majority of the guardians of the legend resent hotly any interpretation which is not in harmony with the tradition. How much this attitude has deterred scholars from writing frankly about Lincoln it is hard to say, but more than one has been heard to declare that he did not feel free to treat the Emancipator as objectively as he could other historical personages. For one or another of these reasons, then, the "scientific" historians have been disinclined to undertake the definitive biography of Lincoln.

Since it is so difficult for any writer on Lincoln to escape entirely from the influence of the legend, it may be well to consider for a moment that extraordinary phenomenon. Its more extreme form, termed the Lincoln *myth,* should be eliminated from consideration because it is not involved in the present discussion. This myth was of popular origin and arose in the years immediately after his death when very little was known of the martyred President's early life. It was the product of emotion and religious mysticism; and it followed in part the pattern of ancient folklore in ascribing to the folk-hero the attributes of divinity, in part an imitation of the story of Christ. The boy Abraham was born in poverty and obscurity, the son of a carpenter; he exhibited in youth the divine qualities of tenderness and mercy toward all living things; under guidance of his Heavenly Father he rose to the leadership of his people at the time of their greatest calamity; in humility and loving-kindness, but with inflexible will, he carried out his divine mission to save his country and free millions of bondmen; and then, because God willed a precious sacrifice in atonement for the people's sins, he died for them. Thus flowered the myth in Christ-like parallel. Had it risen at a time before printing was common, it might have converted Lincoln into a veritable divinity, or at least a prophet or saint; but it has faded from existence as more exact information about the man has become generally available.

The legend, as it exists today, derives from a higher intellectual level. While it lacks the element of religious mysticism as exhibited in the Christ-parallel, its core is a faith in Lincoln's essential perfection in wisdom and conduct both as a statesman and a human being and a conviction that he is immune to any but the most sympathetic criticism. In its broader aspects it pictures Lincoln as the one indispensable leader of his time without whose guidance the American Union, the great exemplar to the world of democracy, could not have been saved nor the four millions of helpless slaves made free, and on the personal side as the great man who was humble, patient, and kind, warmly human, loving and charitable even to his enemies, brooding over the wrongs of mankind, and in the end dying a martyr to a supremely righteous cause. Moreover, he was typically American, rooted in the soil of the New World, drawing his strength of mind and character from the democracy of the mid-continent, rising in the American way from obscurity to greatness.

Although in the building of this legend there has been a marked tendency to ignore many incidents in his personal history, to eliminate or explain away errors, to give him sole credit for accomplishments which realistic historians hold he must share with others, it is at once apparent that there is a factual basis for much of this conception of Abraham Lincoln. In the increasing knowledge and perspective of three-quarters of a century, there can be no doubt that he was one of the most remarkable of our great men. But because the theory of his perfection is based upon selected facts, it is open to criticism. The devotees of the legend do not like to be reminded of any human weakness in their hero or of the cool, shrewd calculation and political maneuvering by which he advanced step by step along the road of his ambition. Some, too honest to deny established facts, nevertheless insist that such actions as may seem regrettable were "not in character" and therefore were but casual and unimportant aberrations, or else that they were, at the particular juncture, necessary to the accomplishment of noble ends. This is not to say that all of Lincoln's admirers have been uncritical of some features of the legend. The late Reverend W. E. Barton, for instance, devoted many years to a critical examination of all the available sources of the Lincoln story and proved that many of the most cherished incidents were mere inventions. Such men are right in holding that a rigorously objective study of his whole career is unlikely to demote the Civil War president from the high place he holds among the great men of America or, for that matter, of the world.

Notwithstanding the criticisms to which it has been occasionally subjected, the vitality of the legend is impressive. How does one explain this perennial interest in Abraham Lincoln and its proliferation into thousands of books, pamphlets, essays, poems, and stories, into references to him as an authority on countless disturbing questions, in appeals to his name on behalf of all sorts of causes? Here was a man who was chosen by a political party as its presidential nominee partly through the manipulations of his political managers, partly because he seemed the "most available" candidate, a man who later as President was regarded by many of his closest associates as too weak and blundering for his task, whom many of his own party would gladly have set aside long before the end of his first term if they could have found a way to do so without damaging their own program, who has now become in the popular mind the embodiment of all the distinctively American virtues, the political seer and leader without parallel and above criticism. Of all the great Americans only Washington may challenge his claim to pre-eminence; but even the great Virginian, though a towering, mountainous figure, seems cold and remote by comparison and evokes no such warmth of popular feeling as does the lanky, homely, sad-faced "rail-splitter" who is believed to have saved and cleansed the nation which Washington had founded. If among native Southerners he still ranks below Washington and Robert Lee, even these people, whose fathers and mothers once hated him for what they had suffered, freely express their admiration for the abilities and character of Abraham Lincoln.

What is the secret of this interest which Lincoln arouses in all classes and conditions of men? It is not enough to say, as some have done, that because he was its first great leader and a successful war president he was popularized by the Republican party, for those astute politicians would never have invoked the shade of the Emancipator if his name had not already been potent with the voters. Was it because he was both wise and good? We have had other great men, even presidents, who possessed both these qualities, but with the exception of Washington none of them holds so high a place in our national pantheon. Somewhat nearer the truth is the theory that his dramatic death, just at the moment of his final triumph when his followers were rejoicing at the end of the four terrible years of war, bestowed the crowning glory of martyrdom and aroused in the minds of a sentimental people a devotion which they might never have felt in like measure had he lived out the natural span of years. But certainly his murder was not enough in itself to account for his apotheosis, since neither Garfield nor McKinley has attained any like status through assassination. While the true explanation of the universal appeal of Lincoln must take account of the special circumstances in which he was placed, for the most part it must be sought for in the peculiar qualities of the man himself. To explain Lincoln and the Lincoln legend seems to have been the fundamental purpose of Mr. Sandburg's stupendous labors.

His contribution to the story of Lincoln's presidency is by far the longest and most detailed yet produced by a single hand, for the four thick volumes contain more than a million and a quarter of words, with hundreds of photographs, cartoons, and other illustrations. One cannot fail to do obeisance to the industry which has compiled so much material, even though some of it is but quotations set into a pattern. Sandburg's method is to follow the familiar order of events but to enlarge the narrative with countless incidents, conversations, newspaper stories, cartoons, editorials, bits of speeches, anecdotes, reminiscences—many of them trivialities in themselves but all together forming an unforgettable picture.

Thus he begins with the confusions and cross-purposes seething in Washington during the winter following Lincoln's election, the failure of compromise, the first secessions of Southern states and the formation of the Confederacy, the strange spectacle of Lincoln's progress from Springfield to Washington with its crowds and speech-making, the crowds and the excited talk in the Capital, the inauguration and the reactions to it, and the frenzied pressure of office-seekers. The reader follows the crowds, hears the speeches, listens in on conferences and half-believes he was a part of it all. After the Fort Sumter crisis exploded in the faces of the startled people, who found themselves caught in "the vast, whining, snarling chaos" of the Civil War, the author relates in the same meticulous detail how the untried new leader faced his awful responsibilities and how in turn other men, and women, reacted to his highly individual conduct of his office.

The story is not always coherent, for much of it is about other persons than Lincoln himself and describes events in which he had no part. In the early portions especially it has something of the confusion and disorder of the time. While the reader is taken most often to the White House to watch Lincoln at work, hear his conversations with his secretaries, and listen to cabinet meetings, or to watch official and unofficial visitors come, have their say, hear Lincoln's replies in various moods, and go their respective ways with all sorts of impressions in their heads, he is also conducted to the halls of Congress, to editorial sanctums, city streets, farm homes, to marching and fighting armies, to ghastly hospitals, and even into the Confederacy. Everywhere are men and women talking of some aspect of the war and commending or condemning the doings of "Old Abe." All this is unlike the ordinary biography: it is more like a vast moving picture with thousands of actors passing in and out, back and forth across half a continent for a stage, but with one central figure who gradually draws the confusion into a certain order, gives it meaning and direction and finally dominates every scene. Little by little, almost imperceptibly, the personality of Lincoln changes too.

Losing nothing of his native shrewdness or his deft political touch, he seems to rid himself of the awkward manners which dismayed so many early White House visitors, to lay aside more and more of the crude jocosity which shocked others, and to reveal a deeper and profounder spiritual quality as he approached the ultimate tragedy awaiting him.

For a study of this character it was necessary to search out sources of information that most biographers neglect or pass over. In addition to all the available writings of Lincoln himself, Sandburg has examined the letters, diaries, reminiscences, and biographies of his contemporaries, newspaper files, sermons, debates in Congress, and even roll calls on bills in which Lincoln was interested, as well as a miscellany of other materials. That enormous product of the Government Printing Office, *The War of the Rebellion: Official Records of the Union and Confederate Armies* in 130 volumes, seems to have proved too formidable for more than a casual examination. In the "Foreword" to his first volume the author has indicated the wide scope of his sources, but in the body of the work he makes no footnote citations. The specialist will recognize the source in most cases but not in all, and will frequently be irritated by his inability to check doubtful statements. The student who may wish to use the work as a guide to unfamiliar materials will find that even when they are indicated in the text he must search a long time for the precise location of the evidence. The general reader, on the other hand, will not miss the footnotes and can surrender himself to the pleasure of reading with few doubts to bother him. Although Mr. Sandburg has evidently kept a wary eye on the specialists, he has written primarily for the much larger and less scholarly group who read books for enjoyment and who do most of the buying.

In the use of the materials he has collected Sandburg has clearly taken great pains to sift out what was true or probably true from the conflicting evidence. When he is skeptical of some quoted statement he warns the reader against accepting it; at other times he flatly denies some contemporary's report of what Lincoln had said or done on a certain occasion, either because it does not fit into the situation or because it is not in keeping with Lincoln's known character. He even accepts the conclusions of other students who have disproved some of the myths most cherished by Lincoln's admirers, such as that about the spectacular dash out to McClellan's camp to pardon the condemned soldier, William Scott. The worst that can be said of his handling of direct evidence is that he either presents the interpretation most favorable to Lincoln when the meaning is in doubt or else refrains entirely from comment when the obvious interpretation must reflect unfavorably upon his hero. It must be remembered, however, that two or three equally honest and equally able investigators may present as many different interpretations of the same evidence, and if Mr. Sandburg's subjective sympathies influence his judgment he has plenty of company among the historians.

Although his presentation of Lincoln as a political leader and statesman conforms very closely with the legend of the great man's nearly infallible wisdom, he is too intelligent and too forthright to assume that "Honest Abe" could never be wrong. But when he must record an action or an incident hard to square with this thesis of perfection, he may point out that larger issues were at stake or that Lincoln was led astray by trusted friends. Such a case, apparently, was the strange affair in which Leonard Swett persuaded the President to issue an order by means of which Swett and his friends planned to steal the rich New Almaden Mine in California. The order was withdrawn after violent protests came in from California. In other instances, such as Lincoln's interference to protect contractors who had defrauded the Government or traders who had illegally sent supplies into the Confederate lines in exchange for cotton or tobacco, the blame is placed upon insistent politicians who pressed purely political considerations upon the harassed President. There are also recorded instances in which Father Abraham was pretty evidently playing politics, when even his extraordinary deftness could not conceal the fact that he was playing the game for all there was in it. Sandburg sees the move and recognizes it for what it was. Sometimes he explains that it was a necessary means to a more important end; sometimes he merely states the facts, seems to wonder at them, makes no comment.

Many of Lincoln's admirers dislike to admit that in the ordinary sense of the word he was a politician at all. They prefer to believe that his was too lofty a spirit to be concerned with such sordid matters as fishing for votes and maneuvering for place, or else they think of him as one who engaged in politics not for personal or party advantage but solely to carry out great public or humanitarian programs, such as the preservation of the Union or the emancipation of the slaves. They forget that these objectives were not a part of his original design, that they arose after he was elected to the presidency and were in a measure forced upon him.[184] They fail to consider that in a democracy like ours a man must first become a successful politician before he has a chance to qualify as a statesman. Lincoln himself would never have denied that he was a politician; on the contrary, he would probably have been surprised and pained to learn that the distinction was withheld from him, for he delighted in his skill as a political manager and felt no squeamishness about the business of political trading for support. The truth is that he loved the game and made no pretense that he did not. But the fact that he had been engaged in

---

[184] COMPILER'S NOTE: Black scholar, Lerone Bennett, Jr. wrote a great book entitled *Forced into Glory: Abraham Lincoln's White Dream* (Chicago: Johnson Publishing Company, 2000). Bennett excoriates Lincoln as a fraud who had, as Ramsdell said, issues forced on him by history that made him look good in retrospect. Bennett pointed out that Lincoln used the "n" word more than the grand wizard of the KKK and, according to Bennett, all of Lincoln's life, he wanted to ship black people back to Africa or into a region in which they could live. Bennett is right. Lincoln's Preliminary Emancipation Proclamation, signed on September 22, 1862, just three months before the actual Emancipation Proclamation, states: "I, Abraham Lincoln, President of the United States of America, and Commander-in-Chief of the Army and Navy thereof, do hereby proclaim and declare that **hereafter, as heretofore**, the war will be prosecuted for the object of practically restoring the constitutional relation between the United States and, each of the [seceded] states, . . . ". It was always Union for Lincoln, NOT ending slavery, because Northern wealth and power were totally dependent on the Union: on shipping Southern cotton and manufacturing for the South. The Preliminary Emancipation Proclamation goes on to say that efforts would continue to recolonize black people out of the United States and back to Africa or into a place that they could live. (Emphasis added.)

politics all his mature life, partly because he liked it and partly because he was fired by an ambition for promotion and fame, need not carry the implication that, after he had reached a position of leadership and responsibility, he would be any the less concerned with the higher obligations of statecraft. Any just estimate of Lincoln the statesman must also take into consideration Lincoln the politician. Sandburg seems aware of this, but he also seems less interested in the politician than in the statesman and less in the statesman than in the personality of the man.

This absorbing interest in the personality of Lincoln is evident on almost every page of the four volumes. Again and again we are brought back from other matters to some apparently trivial thing that Lincoln did, some bit of conversation that throws a light upon one or another facet of his personality or character. The longest chapter in all the four volumes, "The Man in the White House," one hundred and twenty-six pages, consists almost entirely of these little incidents. The reader is taken as a sort of disembodied witness into Lincoln's presence day after day to hear him talk familiarly with his secretaries or with old friends, discuss problems with his cabinet, deal with a miscellany of politicians, army officers, delegations of various kinds, visitors of all sorts and degrees who come on different missions but usually to get something for themselves. We watch the odd gestures, listen to the shrewd, pat comments, the jokes, and anecdotes delivered with loud laughter, perhaps to the reading of a short, carefully prepared paper which leaves nothing more to be said.

Sometimes the tall, sad-faced man is plunged into

dejection by bad news from the army; again, faced by a tearful wife or mother, he pens a little note that is to save some errant soldier from the rigorous penalty of military law. There is much stress on the little kindnesses. Now and then, but not often, he becomes angry and breaks out with a harsh rebuke. More often he manages a sly, shrewd maneuver to turn the tables upon some opponent. Occasionally, scenting political trouble in the .army, he sits down to write one of those careful, equivocally friendly letters to some important personage too powerful to be offended just yet, such as John A. McClernand or Benjamin F. Butler. The net result is a living portrait of an intensely human person who, though not without faults, seems filled with all the homely and essential virtues.

Necessarily there is much about Lincoln's relations with other persons of importance and influence. Some of these people were very difficult, especially those who felt that they owed nothing to the President. But of Lincoln's tactful handling of such men as Sumner, Chase, Stanton, and others there is abundant evidence. Chase was cajoled into staying in the cabinet until his presidential boomlet had collapsed and Lincoln himself safely renominated in 1864, and then was got rid of by a sudden, almost brutal, strategem. Later Chase and his friends were placated by his appointment to the headship of the Supreme Court. The secretive method by which Andrew Johnson was substituted for Hannibal Hamlin as Lincoln's running mate in 1864, in its factual details, is still one of the mysteries, but there seems little doubt that Lincoln's hand was in it. But why the amazing offer, which Sandburg credits, of the vice-presidential nomination to the egregious Ben Butler? The story is liberally sprinkled with evidences of Lincoln's skill in handling men and in the somewhat devious methods of politics. Whether his customary silence in the face of criticism derived from simple, patient self-abnegation, as is so generally claimed, or whether it was dictated by shrewd, secretive caution against stirring up more opposition until time should work a favorable change, is not very clear. Possibly both explanations in varying degrees would answer; but probably the full truth will never be known. Sandburg seems to take a sort of sardonic delight in quoting the criticisms, some of them really venomous, which were showered upon the head of the worried President, not only by the "copperheads," but quite as fiercely by the radicals of his own party—by members of the Senate and House, editors, abolition leaders, ministers of the Gospel—as if he wishes to help them set down the record of their own stupid inability to appreciate their leader, and to contrast their vindictiveness with Lincoln's magnanimity. Against the attitude of these men who assumed to be the guides of public opinion, he sets the increasing confidence and affection of the common people for their President. These people, whom Lincoln understood thoroughly because he had long lived among them and studied them, liked the little stories they read of his kindness to private soldiers and to the poor people who sought him out in the crowded White House.

As for the critics, the reader cannot fail to see that there was much in the outward manner of this frontier President, especially in the early months of the war, to arouse impatience and distrust in the minds of many really earnest and patriotic men and women whom duty or courtesy called to the White House. His unprepossessing appearance, his unpolished social manner, his breezy informality, his frequent lapses into a sort of clownish jocosity when the whole North was depressed by defeat, seemed extremely out of place in one of his official station and grave responsibilities. It is small wonder that those who saw only this aspect of the man who held so much of their fate in his hands should go away with the troubled conviction that he did not measure up to the requirements of his position. No former President, whatever his other defects, bad ever failed to maintain an appropriately dignified front. They could not see that this man was without pretentiousness, that he had so much confidence in himself that he never felt the need to hide behind a mask of pompous dignity. His closer friends and intimates knew him better, and in time many of the critics learned to respect him; but the intensely self-righteous men, like Wendell Phillips, never learned. But even among the common people there was more opposition to Lincoln and his policies than Sandburg indicates. This fact is shown by the elections of 1862 and 1864. In the former year five important Northern states, one of them Illinois, were swept by the Democrats. In 1864, though Lincoln carried all but three states, McClellan received forty-five per cent of the total vote. With that many votes cast against him, it is evident that the popularity of Lincoln, even in the North, was nothing like so great during the war as it was to become in post-war times.

In spite of the vast accumulation of details which have gone into this work, the author still leaves some questions unanswered. For instance, on the disputed question whether Lincoln deliberately provoked the

Confederates into firing on Fort Sumter in order to impose upon them the onus of beginning a war which he thought necessary, Sandburg details most of the evidence but refrains from stating an opinion. For some reason he makes no mention of the entry in Orville Browning's diary in which Lincoln is indirectly quoted as claiming that he had planned for the fort to be attacked and "the plan succeeded."

No one can reasonably expect that four volumes the size of these could be written without an occasional slip, but it is hard to see how some of the factual errors sprinkled throughout these pages could have escaped detection. It will be sufficient for the purpose to select a few from the first volume. The African slave trade was not outlawed by the Constitution of the United States (p. 5) but by an act of Congress of 1807. Kansas had not, when made a state in January, 1861, been knocking for admission for twelve years (p. 14), nor had it always been the South that said "No." Senator Douglas had not voted to buy Florida from Spain (p. 15), since in 1819 he was a child of only six years. It is not true that the South Carolinians would not let Major Anderson's garrison at Fort Sumter have anything to eat (p. 188) at the time Anderson is alleged to have so written to Secretary Holt, about March 1, 1861. From January 31 at latest to April 7, five days before the attack, Anderson had been permitted to purchase both meat and vegetables in Charleston. This Anderson letter, by the way, disappeared soon after it reached Washington and it is not likely that Mr. Sandburg has ever seen it. Texas is listed in place of Virginia in the second group of seceding states (p. 237). The paragraph concerning Jefferson Davis' secession activities (p. 238) and his alleged opposition to the Crittenden Compromise is a tissue of errors. The whole story of the offer of the presidency of the Confederacy to Alexander H. Stephens (p. 255) and his declination is a myth. Stephens did not say in his famous "corner stone" speech at Savannah that the Negro was "subhuman" (p. 256). What he did say was that the Negro was not the equal of the white man. The account of the Confederate law of May 21, 1861, concerning Southern debts due to Northerners as alien enemies (p. 261), is badly garbled. Louis T. Wigfall, the Texas senator, is said in one place (p. 4) to have "winged eight men in duels" and later, in Volume IV (p. 345) to have "killed eight men in duels." Wigfall had fought only two duels and he killed nobody in either. There are many other errors. Mr. Sandburg knows little about the antebellum South and the Confederacy and understands little of what he knows. Except for such established characters as Lee and Jackson, he generally contrives to put something less than the best construction upon the actions and motives of Southerners. He also displays much animus toward George B. McClellan, who certainly had his faults but who was far more generous toward his enemies in official Washington than they were toward him. In fact, Sandburg seems to distrust all whom Lincoln distrusted, whether rightly or not. Nearly all Northern Democrats of the period, except those who came over to the support of Lincoln's administration, he suspects of disloyalty. His treatment of Clement L. Vallandigham is edged with both ridicule and depreciation. Far from merely "pretending candor" the out-spoken Ohio Democrat was, if anything, too candid for his own good, and he was not sympathetic toward secession.

One would expect of Carl Sandburg, poet-journalist of distinction, writing of a high literary quality. Brilliant passages there are, and coruscating flashes, but the quality is uneven. Much of the writing is mere quotation, direct or indirect, sometimes thrown together rather loosely, sometimes cemented together by the author's comments. There is no such sifting, analysis, and conclusion, expressed in the writer's own language, as one finds in Freeman's *R. E. Lee*. Some of it is really dull; but it is not Sandburg who is dull, but his quotations. Among the best things are his sketches of personalities. He has the knack of hitting off a man in a phrase: of Charles Sumner, "The winds of doctrine roared in the caverns of his mind"; Thaddeus Stevens was "a gnarled thorn-tree of a man"; Robert J. Walker was "a somewhat shriveled looking little man, well-spoken, nervous, dyspeptic, adroit, shrewd rather than subtle"; Horace Greeley was "just a little diaphanous: people could see through him—and then again they could not"; and Ben Butler "could strut sitting down." There is also humor, as in the suggestion that General Burnside's bowel trouble was one of the imponderables of the war. One could select from these volumes a whole gallery of vivid portraits.

As the story moves on into the last two years of the war, there is an ever increasing subtle emphasis upon the spiritual growth of Lincoln. And as the final scene approaches, a sort of orchestral tragic *motif* begins to throb through the notes of triumph at the dawn of peace. The account of the assassination is full, vivid, moving. In the chapter entitled "A Tree is Best Measured when its Down" are gathered an extraordinary collection of contemporary estimates of Lincoln from all over the world, most of them written while the nation was still under the shock of his murder. Then follows the final chapter, a funeral dirge of sweeping, melodious grief, the highest point in the literary quality of the work. These four volumes were not intended to and do not provide an adequate history of the Civil War; nor, although they bring together nearly all the factual data available about Abraham Lincoln during the years of his presidency, do they constitute the definitive biography. Perhaps such a work will not be possible until after 1947 when the mass of Lincoln papers in the Library of Congress shall be opened to inspection. But the definitive biography, if ever there is to be one, must be done by some one with a broader historical training than Carl Sandburg possesses, by one who knows more of the complex issues which brought on the sectional conflict as well as those of the war itself, by one who is able to free himself from the power of the legends which have grown up about Lincoln. That will be no easy task. But our definitive biographer will be fortunate if he can match in sheer interest the four volumes Carl Sandburg has given us.

The Southern Heritage[185]

by

Charles W. Ramsdell

For four long, weary years they fought an unequal and inherently hopeless struggle for independence, enduring what no civilized people had suffered since Louis XIV ravaged the Palatinate. In the end they were conquered, their slaves freed and their social system overturned, their lands laid waste and left almost worthless, their mills, factories and railroads destroyed or weakened beyond immediate repair, their banks and accumulated credits wiped out, their families impoverished, the flower of their youth dead or crippled.

I.

The dominant factor in the history of the South has been its preoccupation with agriculture and especially with the great staple crops of cotton, tobacco and sugar. Despite

---

[185] Charles W. Ramsdell, "The Southern Heritage." In *Culture in the South*, William T. Couch, ed. (Chapel Hill: University of North Carolina Press, 1934), 1-23. Article and citation are verbatim.

the rapid growth of cities and the development of mechanic industries in recent years, the bulk of the population is still primarily concerned with the cultivation of the soil.[186] The economic and social status of the agricultural class, therefore, has largely determined both the condition of southern life and the patterns of thinking. But the regional situation has long been complicated by other factors. In the first place there is the race problem, which was deeply involved in the southern defense of slavery and, far from being eliminated, was made more difficult by the emancipation of the blacks. Another factor which has had much influence upon the regional attitude, one which developed before 1860 and was emphasized by conditions after Appomattox, was the consciousness that the South was a sectional minority, relegated to a subordinate position in national affairs, and that it must be constantly on the defensive against the policies of the rich and powerful North. And of far-reaching importance, in both its social and its psychological consequences, was the blighting poverty which was inflicted upon the whole region by the Civil War and the ensuing decline of agriculture. Even now, two-thirds of a century after the debacle of 1865, although the South has recovered much lost ground, it has not regained a position of importance in national life comparable to that which before 1850 was the foundation of its sectional pride .

The differences between North and South, based on soil and climate, began far back in colonial days, but they were not great enough at that time to cause friction. Both sections were concerned chiefly, though not in the same

---

[186] Ramsdell is speaking of the 1930s.

degree, with agriculture and trade. It was not until the opening years of the nineteenth century that their interests began sharply to diverge. It was then that the industrial revolution began to change the economic organization of New England and the middle states, while through its textile machinery and the cotton gin it suddenly made raw cotton a staple of the world market and turned the agriculture of the lower South in a new, profitable and fatal direction.

While mills and factories multiplied and capital concentrated in the North Atlantic states, eager farmers and planters along the coast and far into the back country of the Carolinas and Georgia were growing cotton and dreaming of larger estates and greater profits. High prices for the raw staple made fortunes for successful planters and the "cotton craze" spread to all classes. After the War of 1812 a steady stream of settlers moved into the new Gulf states, cleared the forested lands along the rivers, and by 1830 were producing over half of the total crop. The development of this new southwest was phenomenal, but it is easily explained. Cotton was not only a reliable cash crop, since it sold readily in the world market, it was also normally the most profitable of all crops on good soil. Above all, it was well suited to large-scale production with Negro slave labor. Slaves, in fact, were essential to large planting units; for the task of clearing the thickly wooded river lands and carrying on the routine of new plantations required constant labor. A steady and adequate supply of free white labor could not be procured on this frontier where even the poorest white man could always get land to cultivate for himself. The large landowner must have Negroes or leave his fields unplanted. The successful cotton grower found that the most promising investment for the profits from his crop was in more land and more labor. It was, in fact, almost the only one open to him. Thus the plantation system continued to expand and the lower South was committed to cotton and to slave labor by economic forces beyond its control. The border states were less affected by these momentous changes. Cotton growing made slight headway in North Carolina and Tennessee, and in Virginia, Kentucky and Missouri practically none at all. In these states farmers grew tobacco or hemp, grain and live-stock. They used Negro slaves, but in no such numbers as did the cotton planters to whom they sold their surplus of Negroes. In this upper country, and also in the piedmont and the mountain valleys of the lower South, there was a more diversified industry than in the cotton belt. Difficulties of transportation increased the cost of imported goods and thus made opportunities for local white artisans. Small capitalists erected flour and grist mills where water power was available; others opened coal and iron mines and salt springs; and a few built small yarn and cloth mills. But these were generally local enterprises to supply the local market, and they held their business chiefly because the absence of cheap transportation protected them from outside competition. Richmond, Petersburg, and Louisville became small industrial centers, but only because of peculiar local advantages. Meanwhile these enterprises had little effect upon the general trend of southern life, for agriculture and the commerce which it fostered remained the chief concern of the mass of the southern people.

As late as 1850 the country was thinly settled and

retained many of the characteristics of the frontier. Most well-to-do farmers still lived in log houses. Bad roads and unbridged streams made travel and communication difficult, while heavy woods about the cleared fields deepened the sense of isolation. The small towns, with few exceptions, served merely as shipping points and offered little to attract residents. The southerner preferred to live in the country. Although the larger cities maintained daily newspapers, the mass of the people read only little four-page weeklies. In rural communities schools were hard to establish and maintain. Farmers' children went for two or three months in the year, if at all, to local schools which were supported by small tuition fees and the "boarding around" of the teacher. The older and more ambitious boys sometimes found opportunity to attend one of the numerous private academies. The children of the wealthier planters had tutors and access to libraries of English classics. This was the older system; but in the eighteen-thirties and forties it was beginning to give way to the pressure for tax-supported public schools, and before 1860 every southern state had established some form of public school system. State supported colleges antedated the common schools and were better provided for. In fact, the best of the southern colleges were doubtless as good as the best in the North.

Like their countrymen of other sections, the mass of the southern people were sincerely religious. Except in the French parishes of Louisiana, Protestantism was everywhere in control. The older aristocratic families maintained the prestige of the Episcopal Church which had an influence far out of proportion to the number of its members; but for ordinary people the Presbyterian,

Methodist and Baptist churches had a stronger attraction. The preachers of these evangelical churches, by flaming appeals to the fears and emotions of the common people and by emphasizing their individual importance in the sight of God, had gathered them into the fold by the thousands. The members of these churches accepted the Bible as the literal word of God, the infallible guide for human conduct, the fountain of all wisdom. Since each church had its own interpretation of God's plan for the salvation of souls, it was important that doctrine be sound; therefore theological controversies abounded and rivaled politics in their intensity. But the clergy of all the evangelical churches worked in kindred spirit to impose puritanical ideals upon their people and to root out the tolerant deistic cult of Thomas Jefferson and Thomas Cooper. They sought to control the education of the youth not only by erecting sectarian colleges but by exercising a very direct influence over the state universities.

His political party, like his church, gratified the individual's sense of his own importance, but the party had a more universal appeal than the church. The average voter formed his opinions rather less from reading than from listening to political speakers;. so that public debate became a system of political education and oratory one of the fine arts. Young politicians acquired ample schooling in party organization and management in the sharp conflicts between Democrats and Whigs over a multitude of issues, local, state and federal, or between rival leaders of the same party. The leaders were usually men who had enough means to enable them to pursue a public career as an avocation, most often as an adjunct to the law. As a rule, public officials seem to have been free from the taint of graft; at any rate, there were few scandals. State and county governments were simple and economical, holding as far as possible to the old agrarian doctrine, "that government is best which governs least."

Before 1830 the phenomenal spread of cotton growing into the southwest had brought about a gradual, if fluctuating, decline of prices which began to pinch the planters on the thinner soils of the eastern cotton belt, especially in South Carolina. At the same time the rapid accumulation of banking capital in the industrial and commercial centers of the northeast had enabled northern bankers and brokers to obtain control of the marketing of the crop. Normally the planter shipped his cotton to a commission house or "factor" at the nearest seaport, and from him purchased his yearly supplies. Often the factor was merely an agent for a northern firm which sold the cotton in New England or Liverpool, and subtracted liberally from the proceeds for wharfage, freight, insurance, storage, drayage, and commissions. Whether crops were good or bad, prices high or low, the factors and bankers were able to take their own profits. But the planter found it harder and harder to make a profit from his crops, and the declining trade of Charleston reflected his condition.[187]

---

[187] COMPILER'S NOTE: This HUGE profit made by Northerners every step of the way in marketing Southern commodities, especially cotton, is why the North could not let the South go in peace in 1861 as the South wanted and expected. Not only would Northerners lose this enormous profit, Southerners would then make that profit themselves off their own hard work and they would begin to accumulate capital for their own industrial development. They would also trade freely and directly with Europe, as they had always

---

wanted, thus devastating the Northern shipping industry and further cutting Northern profits. This HUGE amount of wealth constantly flowing into the North does not even include gargantuan sums the North made off the federal government beyond the tariff. Northerners had monopoly protection for many industries meaning there was no market competition so Southerners had to pay astronomically jacked up Northern prices; and Northerners benefited from payments made directly from the federal treasury into Northern pockets in the form of bounties and subsidies for many industries. Henry L. Benning stated that the amount was $85,000,000 per year — a gigantic sum in 1860 — that flowed out of the South and into the North for which the South received not a thing. The greatest economist of the nineteenth, Thomas Prentice Kettell, in his brilliant book, *Southern Wealth and Northern Profits, as Exhibited in Statistical Facts and Official Figures: Showing the Necessity of Union to the Future prosperity and Welfare of the Republic* (New York: Geo. W. & John A. Wood, 1860), documents and makes it crystal clear that Southerners were producing the wealth of the United States with cotton and other Southern commodities but Northerners were taking all the profits. The injustice of this situation was greatly magnified because 3/4ths of the tax money in the federal treasure came from the South, but 3/4ths of the treasury was being spent in the North. In *The Address of the People of South Carolina, Assembled in Convention, to the People of the Slaveholding States of the United States* (December, 1860), South Carolinians stated that "The Southern States now stand exactly in the same position towards the Northern States that the Colonies did towards Great Britain. The Northern States, having the majority in Congress, claim the same power of omnipotence in legislation as the British Parliament." They went on: "The people of the Southern States are not only taxed for the benefit of the Northern States, but after the taxes are collected, three-fourths of them are expended at the North." No wonder dishonest historians want to keep a red herring light on slavery. The economic theft going on in 1861 was a far more powerful reason for the South's secession and desire for independence, especially when it is proven beyond

Under such circumstances the protective tariff, which had risen steadily since 1816 and had increased the cost of living, was an extreme irritation. The distress in South Carolina was so great by 1832 that her people struck at the tariff through nullification. Although the other southern states refused to endorse the South Carolina remedy, the compromise tariff of 1833 relieved the strain. Andrew Jackson had stopped federal aid of internal improvements in 1832 and destroyed the Bank of the United States in 1836. Agriculture had won the first political battle and it was to win others with the low tariffs of 1846 and 1857; but it had not won the campaign. The northern merchant and banker were still taking their accustomed profits from southern trade.

The general economic depression of the late thirties induced a more extensive search for causes and remedies. While some southern planters followed the lead of Edmund Ruffin of Virginia in instituting "agricultural reform" by fertilizing and terracing their fields and rotating their crops, others sought more direct means of breaking their vassalage to northern capital. They had not failed to note that from two-thirds to three-fourths of the total exports of the United States were the products of southern agriculture and that an even larger proportion of

---

the shadow of a doubt that the North did not go to war to free the slaves or end slavery. Their War Aims Resolution, the Corwin Amendment, the fact that five slave states fought for the North throughout the war, the Preliminary Emancipation Proclamation, Lincoln's voluminous statements, etc. ALL make it clear that the North was fighting to preserve the Union, not end slavery. Now we know why. Their enormous wealth and power were dependent on the South and the Union. (Emphasis added.)

the imports came to northern harbors. Since it was evident that northern merchants and bankers paid their accounts in Europe with the country's exports, the planters easily concluded that northern wealth grew largely from the profits of handling southern trade. These profits might be saved to the South. Led by southern merchants, they proposed to cut loose from New York, Philadelphia, and Boston by establishing direct trade with European markets through new lines of steamships and to strengthen this trade by railroad connections between the South Atlantic ports and the Mississippi and Ohio valleys. The great scheme failed because the scarcity of available capital delayed the building of the railroads and because no steamship line could be induced to send ships regularly where freights were seasonal. Northern capital continued to skim the profits from the southern cotton trade.

During the early forties, when the price of cotton fell to ruinous levels, there was much talk of establishing cotton mills in the cotton country. William Gregg of South Carolina and a few other industrial pioneers had pointed the way to success in this direction, and small factories arose along the eastern piedmont from Richmond to Augusta; but because of insufficient capital, inexperience and bad management there were enough failures to discourage investors. When the price of raw cotton rose again men with spare capital turned back to planting—something which they understood and could manage themselves. Nevertheless, the manufacture of yarns and the coarser fabrics made headway, and had they not been set back by the Civil War, the southern mills might have dominated the industry before the end of the century. In the upper South, Richmond had developed a nucleus of tobacco factories, flour mills, and iron manufactures, while Louisville had for some time been one of the industrial centers of the Ohio valley. But the mass of the southern people continued to till the soil and to exchange at heavy cost their raw produce for northern manufactured goods. As late as 1860 more than ninety per cent of the manufactured products of the United States, exclusive of such things as lumber, turpentine, flour and meal, were fashioned north of the Potomac. As better times returned in 1850 the planters recovered their confidence in the security of agriculture. The Walker Tariff of 1846 had lowered the cost of manufactured products and southern cotton was in active demand in the markets of Liverpool. There is little doubt that their renewed prosperity was a large factor in winning their acceptance of the great Compromise of that year. What they did not see, or else were unwilling to admit, was that a system of exchanging raw agricultural products for imported manufactures bound the South to a *colonial* economy, and that the results would be the same whether they bought their goods in the North or in Great Britain. Although they exultantly proclaimed with Senator Hammond of South Carolina that "Cotton is King!" they gave evidence of latent uneasiness. They were still irritated that there was no escaping the tribute levied by northern capital and industry, and a small group of irreconcilables began to insist that economic independence could be obtained only by political separation from the overshadowing North. But the economic arguments of the secessionists had little perceptible effect upon the strong sentimental attachment to the Union. Had it not been for the excitement which had arisen over the slavery controversy, it is unlikely that economic grievances alone would have brought the South to disunion. They certainly played a minor part in bringing on the revolt of 1860-1861.

The rapid development of the anti-slavery movement in the North, at the core of which was a virulent and uncompromising demand for immediate and uncompensated abolition, seemed to the southern people a menace far more dangerous than the aggressions of northern capital. It aroused an emotional storm in the South which is hard for most people nowadays to understand. This feeling cannot be attributed wholly to the fear of the loss of the capital invested in slave labor—although that was naturally an important factor—for some of the most determined opponents of abolitionism were men who owned no slaves themselves. It was, in part, a quite natural anger at the rabid accusations and wholesale abuse heaped by the abolitionists upon all men and women who had any connection with the institution or who defended it. And it was, in very large part, a fear of the consequences, if the crusade should prove successful. It must be remembered that when Garrison began the publication of *The Liberator* in 1831 many thousands of Negroes in the lower South were either natives of Africa or were but one generation removed from savagery. As long as these Negroes could be kept under the discipline of the plantation the danger of racial conflict would be minimized; but should they be excited by abolitionist doctrine the consequences might be appalling. Southern men had not forgotten the horrors of the Haiti massacres nor the more recent Nat Turner insurrection in Virginia. To them the abolitionists were fanatical incendiaries, reckless of the lives of the white women and children of the "black belt." For us to say now that the danger of servile uprising was far less than they thought is beside the point. That they believed the danger real is the important fact. On the economic side the abolitionist program threatened not only the loss of capital invested in slaves but also the shrinkage of land values, since it was commonly believed that the Negro could not be depended upon to do much work after he was freed. And it must be remembered that the lands of the slave belt were, on the whole the most fertile in the South. The sum of the matter is that the abolitionist attack seemed to be aimed not merely at the basis of southern economic life but at the existence of white civilization itself. Confronted with these dangers and compelled to listen to lurid descriptions of themselves as monsters of brutality and iniquity, it is small wonder that southern men lost their tempers and therefore lost the power to think objectively on the subject of slavery. The distinction between the abolitionist creed and the milder and much more general form of northern antislavery opinion, which advocated limitation of the institution rather than, as yet, its destruction, seemed to them less and less important; they were convinced that all antislavery sentiment was working around to the abolitionist position. Thrown upon the defensive, they began elaborating a defense of slavery and of everything southern. As the fatal controversy dragged on, every attack from the North deepened the defense patterns in the southern mind. Southern political history during the two decades before secession cannot be understood without taking into account this conviction of being on the defensive and the extreme sensitiveness which it induced.

Since the foundation of the government the South had been a minority in the Union, but this weakness seemed of little importance until the great sectional issues arose. The great Calhoun had turned all the resources of his mind to the problem of protecting the minority against the potential tyranny of the majority, and had failed to find a practical solution. There was no solution; but southern statesmen erected one defense after another—constitutional limitations on federal power, political alliances, equal representation in the Senate. All failed. They fell back upon the sovereignty of the state, a doctrine which had been asserted at some time in every section of the Union. It was the last refuge short of secession and acquired the authority of a religious dogma. But as the tide of population rolled westward into new lands that could never be southern, while certain of the border slave states showed a declining interest in slaveholding, it became a question whether the sacred principle of state sovereignty could long withstand the pressure of numbers. What would happen when enough free states had been carved out of the West to give the anti-southern forces three-fourths of all the states in the Union—enough to override the last defenses of the South by constitutional amendments? The desperate tactics of the southern politicians after 1850—the demand for unhampered entry of slave property into western territories where climatic conditions effectively barred the institution, the repeal of the Missouri Compromise, which could give no possible practical advantage to their section—however futile they may seem now, were all fundamentally measures of defense. Their effect was to bring into existence immediately the first great sectional party, the Republican, a party which in some form was certain to emerge sooner or later. The "Black Republicans" (a term which sufficiently distinguished them from Jefferson's organization), whose cardinal principle was opposition to the extension of slavery,[188]

---

[188] COMPILER'S NOTE: The Republican Party and Northerners in general did not want slavery in the West because they did not want blacks in the West. Most historians agree on this. The abolitionist movement, ironically, had an anti-black aspect. As Lincoln said in the Lincoln-Douglas Debates, the West was to be reserved for white people from all over the world. Many Western as well as Northern States had laws on the books forbidding free blacks from living there. In Lincoln's Illinois, a black person staying longer than a few days was subject to arrest, incarceration, whipping, etc. Historian Charles P. Roland in *An American Illiad, The Story of the Civil War* (Lexington: University of Kentucky Press, 1991), 3, states: "There was a significant economic dimension in the Northern anti-slavery sentiment" and "a racial factor contributed to the Northern attitude" because "Many Northerners objected to the presence of slavery in their midst, in part, because they objected to the presence of blacks there." Historian David M. Potter in *Lincoln and His Party in the Secession Crisis* (New Haven: Yale University Press, 1942; reprint, New Haven: Yale University Press, 1979), 200, states that Northern anti-slavery was "not in any clear-cut sense a pro-Negro movement but actually had an anti-Negro aspect and was designed to get rid of the Negro." Charles W. Ramsdell in his excellent treatise in this book, "The Natural Limits of Slavery Expansion," argues persuasively that slavery was not expanding into the West and could not work there. His argument is boosted significantly by actual fact from the census. In two large Western territories that had been open to slavery for 10 years, there were only 24 slaves in one (New

professed to harbor no designs against the institution in the states; but they called the abolitionists to their support and counted scores of them among their leaders. Southern men refused to be taken in by platform professions. They knew that Black Republicanism was their implacable enemy.

In October, 1859, John Brown's attempt to head a slave insurrection in Virginia, followed by the northern panegyrics lavished upon that "martyred" incendiary, aroused apprehension and indignation throughout the South. Doubtless many of the secessionists secretly rejoiced that the final issue was brought closer; but the majority of the southern people were troubled by the portent of danger. Would they be safe if the antislavery men should win control of the government at the next election? A year later the election of Lincoln and the refusal of the victorious Republicans to agree to any compromise or satisfactory guaranty of their safety

---

Mexico Territory), and 29 in the other. The expansion of slavery into the West was nothing but a Northern racist political argument in 1860 with Lincoln and the Republican Party appealing to racist white Northerners that if they voted Republican, the West would be theirs. Both sides became rigid over an issue that Republican James G. Blaine stated was "related to an imaginary Negro in an impossible place." Lincoln scholar Richard N. Current in *The Lincoln Nobody Knows* (New York: McGraw-Hill Book Company, Inc., 1958), 95-98, wrote that "Lincoln and his fellow Republicans, in insisting that Congress must prohibit slavery in the West, were dealing with political phantoms." Congress "approved the organization of territorial government for Colorado, Nevada, and Dakota without a prohibition of slavery" because they did not think it necessary. In 1860, there were only two slaves in Kansas and 15 in Nebraska, and that was after being open to slavery for 10 years. Ramsdell was absolutely correct.

convinced them that the time had come for separation and the establishment of a new southern union. In mingled anger, regret, and enthusiasm the cotton states withdrew and set up their Confederacy. The upper South held back until Lincoln's call for troops; then four more states joined to resist the armed invasion from the North.

If the southern people erred grievously in their decision, grievously they answered for it. For four long, weary years they fought an unequal and inherently hopeless struggle for independence, enduring what no civilized people had suffered since Louis XIV ravaged the Palatinate. In the end they were conquered, their slaves freed and their social system overturned, their lands laid waste and left almost worthless, their mills, factories and railroads destroyed or weakened beyond immediate repair, their banks and accumulated credits wiped out, their families impoverished, the flower of their youth dead or crippled. Their conquerors imposed upon this stricken people, in the form of taxes and the repudiation of their bonds, an indemnity heavier than that levied upon France by Imperial Germany in 1871. They faced their condition courageously and sought to build anew upon their broken foundations; but the triumphant North fell under the direction of vindictive leaders who utilized to the utmost the prevailing war psychosis to frustrate every effort of the southern whites to recover something from the general wreck. The orgy of radical reconstruction kindled a new resentment more enduring than that caused by the war itself.

By all odds the bitterest draught held to the lips of the southern whites during reconstruction was their enforced subjection to the political control of their former slaves. Not even the imposition of exorbitant taxes and the plundering of the public treasuries by the carpetbaggers equaled in their eyes this iniquitous humiliation. The purpose of the northern radicals was not so much to uplift, the Negro as to "nationalize" the sectional Republican party and thus perpetuate their own control of the government. The Negroes, of course, were not to blame. They had not sought this power; but their ignorance and credulity, their naive delight in their new importance made them easy tools in the hands of unscrupulous leaders. It was impossible to eliminate the venal carpetbagger without depriving the Negro of his political power; and in the process racial antipathies developed which were to be a source of trouble for years to come. There can be no doubt that violence and cruelty toward Negroes was greater after the freedmen were first given the ballot in 1867 than it ever was under slavery; but it is significant that the lower class of whites—those nearest the Negro's social and economic level—were the most active in these disturbances. As the Negro was gradually retired from politics, racial friction decreased, and an increasing number of southern leaders took the enlightened position that in order to make him a useful citizen, the freedman must be encouraged to acquire an education and property. Despite sporadic outbreaks, millions of whites and blacks were able to work together on friendly terms.

Inevitably, perhaps, this solution of the problem of race relationships found little acceptance in the North where the war had engendered a persistent suspicion of southern motives and where radical Republican "bloody shirt" orators and editors sought to hold their party lines intact by charging southern Democrats with being "as rebellious as in 1861" and with practicing every sort of cruelty upon the "Union-loving freedmen, the wards of the nation." The effect of all this was further to inflame the southern mind and to impede the efforts of enlightened and farseeing men to improve the condition of the Negro as an essential step in the regeneration of the South, since to the less thoughtful majority every plea for the freedman was linked up with the anti-southern propaganda of the hated northern radicals. Again rationalizations about the dangers of "social equality" retarded the solution of the problem of racial adjustments.

Northern publicists continued as they had begun in 1861 to explain the war as "the spawn of those twin devils, treason and slavery." For forty years this theory was elaborated and reiterated until it became accepted in the North as one of the historical verities. Thus the burden which the conqueror always manages to lay upon the spirits of the conquered—the heavy burden of sole responsibility for so much suffering—was loaded upon the southern people. To make it more irritating, southern parents saw school histories put into the hands of their children which discredited their own motives in the desperate struggle they had made to protect their land from invasion and devastation and stigmatized them for rebelling against the moral and philanthropic designs of the righteous North. Defenders of the southern cause were numerous, but they could get little hearing. Southern political leaders, after they had been allowed to return to Congress, could ill afford, for the sake of their northern Democratic allies, to arouse northern susceptibilities by attempting the vindication of the movement for southern independence. Few were inclined to reargue the cause of slavery, now that it was gone; most men, in fact, expressed satisfaction that it was no more. But none were willing to stultify themselves by admitting that their motives had been bad in 1861. While those who remained "unreconstructed" proclaimed with Robert Toombs that they "had nothing to regret but the dead and the failure," the majority turned hopefully to the future and sought to find compensation for their children in the new South which was somehow to rise miraculously from the ashes of the old.

It was expected that recovery would come through agriculture, for cotton brought an abnormally high price at the end of the war. A few good crops would bring large returns and upon this basis other industries could rebuild. The result was disappointing, for cotton growing recovered only slowly during the turmoil of reconstruction. Throughout wide areas where the federal armies had penetrated, houses, barns, gins and fences had been burned, the stock killed or driven away. The land remained, but its market value was less than half that of 1860 and it often had to be mortgaged in order to procure the means for starting anew. Even the districts which had escaped ravage were in a sad state—tools worn out or broken, stock depleted, neither money nor credit procurable except on ruinous terms. Some of the small cotton crop of 1864 remained in the country, but most of it was seized by rapacious federal agents. The acreage planted in 1865 was small and the harvest generally was poor. The government exacted a tax of three cents per pound on ginned cotton. The labor of rebuilding houses, barns, gins, and fences took time and the cost absorbed the small profits as fast as they came. Although the Negroes, after the first year or so, worked better than was expected, they were kept in a continual state of unrest by carpetbag politicians and soon demanded separate fields as tenants in order to be freer of control. In fact, their former masters were usually unable to pay them monthly wages and could offer them only a share of the crop. The result was that the large plantations were divided into smaller tracts, each worked by a Negro cropper on shares, or else were broken up and sold to small white farmers. The slipshod methods of the cropper reduced the output per acre to less than before the war, so that it was not until 1878, thirteen years after Appomattox, that the cotton crop east of the Mississippi equaled that of 1860. By that time prices were again low; and as the yield increased thereafter they declined still further and profits vanished.

When the planter became the landlord of Negro croppers he must furnish them with teams, tools, food, clothing, and shelter while they made the crop. As he seldom had sufficient cash to provide these things, he must turn to the local merchant for credit, pledging his share of the crop as security. Because of the uncertain quality of the security, he could no longer depend upon the distant factor who had been his creditor before the war. The small farmer, also, must buy on credit from the merchant and give a lien upon his crop. The merchant in turn obtained advances from the wholesaler or jobber, and the chain of credit ran back to the northern manufacturer and his banker. The rate of interest upon the credits advanced on the crop, usually concealed in the high prices of the goods sold, ran from twenty-five to seventy-five percent. No ordinary business could endure such a drain; and as the prices of farm products fell, farm owners found themselves hopelessly mired in debt. To attempt diversification of the crop was useless, for the merchant required a large acreage in a "cash crop"—cotton or tobacco—to cover advances of credits or overhanging debts. Hopelessly entangled, thousands of once independent farmers either sold their land or lost it by foreclosure and passed into the status of tenants, or else moved west to Texas to start over again. Farms and plantations were bought up by merchants who then supplied credit directly to their own tenants. Some of the planters moved to town to become merchants themselves. Economic power and prestige passed rapidly from the owner and tiller of the soil to the merchant, banker, and professional man of the town .

As agriculture sank under its burdens, industrialism swept on to increasing power in national affairs. While the southern statesmen, the ablest of the agrarian leaders, were absent from Congress for some fifteen years, the masters of industry had improved their opportunities by obtaining new favors from Congress and by entrenching themselves in the Constitution. They had obtained higher and higher tariffs; they had procured a new national banking system which greatly extended the power of capital over the terms of credit; they had increased the burdens of debtors and the holdings of creditors through the manipulation of federal finances; and they had inserted a "joker" in the fourteenth amendment which effectually prevented the states from controlling the great corporations which did business within their borders. The

Republican party, more and more dominated by the eastern wing, was now the servant of big business; and a large section of the eastern wing of the Democratic party, to which the now "solid South" clung with pathetic faith, was committed to the same political-industrial program. The agriculturalists, the cotton and tobacco growers of the South and the grain growers of the West, divided by the bloody chasm of the war, could not make successful resistance. This condition could not endure without revolt; but successful revolt was made difficult by the peculiar status of southern politics. The grim ten-year struggle of the reconstruction period had centered about the question whether the southern whites would be able to recapture control of their local governments. Under the pressure of the northern radicals and the local carpetbaggers and Negroes, southern antebellum party lines were obliterated and the mass of southern farmers, planters, business and professional men fused into one party, the Democratic. The "solid South" for which Calhoun had worked in vain had arrived—far more compact than at the time of secession. Apprehension of further federal interference in local affairs, especially in behalf of Negro activities in politics, was a powerful cementing force; and for fifteen years more the political leaders who had cleansed their states of the carpetbag governments continued to guide and rule. They were generally men of high character and great popularity, and the widely conceded necessity of maintaining Democratic solidarity seemed to make their position impregnable. But these leaders, largely of the old planting and professional classes or former officers of the Confederate army, were being drawn into ever closer relations with the new business enterprise which was stirring in the South. Conservative because of tradition and business association, convinced believers in the doctrines of *laissez faire,* they saw only danger in governmental interference with private business. And this at a time when symptoms of revolt were sweeping through the masses of impoverished farmers and tenants—revolt against the unseen forces which depressed the prices of agricultural products and increased debts, against the exactions of the money lender and the extortions of the railroads.

The first agrarian revolt came not in the South, however, but among the farmers of the West who organized the "Patrons of Husbandry" or Grange in the late sixties to provide for better agricultural education, to curb the railroads and monopolies, and to devise a scheme for co-operative buying and selling. The order spread into the cotton country during the next decade, but it collapsed before the southern farmers, still struggling with the fag end of reconstruction, awoke to its possibilities. In the late eighties another organization with a similar purpose, the Farmers' Alliance which began in Texas, made better headway in the South; but internal dissensions, fomented in part by the apprehensive business element, dissipated and destroyed it. Both of these organizations had endeavored to avoid partisan politics in order to prevent divisions within their ranks; but when the farmers found that their old political leaders had nothing to offer them, their discontent broke over into political action. In the South it took the form at first of an uprising against the Bourbon leadership within the Democratic party. In 1890 came the Tillman revolution in South Carolina against the silk stocking aristocracy of the towns. James S. Hogg fought his way to the governorship of Texas with a mandate to curb the insolence of the railway barons. The farmers of Georgia staged a revolt against the triumvirate of Gordon, Colquitt, and Joe Brown. The Bourbons everywhere were suddenly on the defensive. In 1892 the People's Party, forming with startling swiftness, attempted to restore the power of the farmer in both state and national affairs. The southern Populists recalled the Negro to the polls, an evidence of their desperation which startled the conservatives and soon led to more stringent restrictions upon the Negroes' use of the ballot. But the Populist movement broke against the solid wall of party regularity and eventually disappeared, leaving behind it a program which was later adopted, in part, by the old parties after it was seen to be harmless. The Democratic party was itself captured by the advocates of free silver coinage, who thought by increasing the supply of money to get better prices for agricultural products and thus lift the burden of their debts. It was the misfortune of the debt-ridden farmer in his fight against the strangling hold of the capitalist creditor to be forced always to the support of "unsound" economic theories or of "unconstitutional" legislation. When he took up the weapons of his antagonist—cooperation and combination, as in the Grange, the Farmers' Alliance, and the Farmers' Union—his own individualism helped to break the organization down.

Defeat, poverty, frustration, the wearying sense of being eternally on the defensive had their inevitable depressing effects upon social and cultural conditions. The tone of southern life was lower than before the war.

The old planter aristocracy, which before 1861 had set the standards of polite society, was broken, scattered, and losing its leadership. It has been the fashion of writers of the forward-looking school to count this as a blessing second only to the destruction of slavery; but before accepting this dictum it may be worth while to inquire what this social order was. While it must be acknowledged that the popular legend of the antebellum southern aristocracy contains many absurdities, no one who has honestly tried to understand the group can escape the conclusion that it had qualities which neither the South nor the nation can well afford to lose. It was not necessarily a class of great wealth, even as wealth was counted in those days, but of enough means to afford travel, the possession of books, and the leisure to read them. Its essential character was in its way of life and its attitude toward life; and these, the result of generations of good breeding and quiet living, were rooted deep in traditions of family honor, public responsibility, self-respect, a contempt for lying and cowardice—in short they were the standards of gentlemen and gentlewomen in every age and country. Not all the aristocracy, of course, lived up to the ideals of their class, but the finer representatives of the group exercised an influence on southern life far out of proportion to their numbers. Submerged by the social and economic revolution which followed the war and by the gradual rise to leadership of new groups with "go-getting" ideals, they did not lose their identity nor wholly change their ways of life, for family tradition is strong in the South, but they no longer determined the standards of social conduct.

Personal and sectional poverty, the realization that

the South no longer shared in the direction of national affairs, everything that emphasized the painful contrast with the past, caused many of the older generation to look back wistfully to the good old days when the South had an integral and important place in the life of the nation. It was easy to slip into an idealization of the old South, to tinge its receding memories with a roseate glow, to picture it as composed principally of the fine old aristocracy whose homes were ever filled with light and laughter and whose prosperous fields were tilled only by happy, loyal and contented slaves. There was enough substance in the tradition to give it value, and it was seized upon by social climbers, especially by women eager for social prestige and by those shallow nuisances, the professional southerners. Partly from motives which were wholly justified and partly as a defense reaction to northern misrepresentation, the Lost Cause was becoming glorified. Justly proud of the records of their desperate valor, survivors of the Confederate armies organized to perpetuate the memories of the War between the States and to commemorate their dead. Hence arose those pathetic[189] monuments which adorn courthouse and capitol lawns throughout the South. And for thirty years an almost indispensable qualification for public office from Virginia to Texas was a good record as a Confederate soldier.

In the years immediately before the war some form of a tax-supported system of public schools had been

---

[189] One of the meanings of the word "pathetic" is "affecting or moving the feelings," and that is how Ramsdell uses it. http://www.dictionary.com/browse/pathetic, accessed March 8, 2017.

established in every southern state. In most states the system was very imperfect; but a few had excellent schools, and in all there was evidence of improvement and growing public interest in the subject. These schools were the fruit of the democratic movement which had gone steadily forward for a quarter of a century, during a time when the South is commonly supposed to have been under the domination of reactionary planters. The war and its aftermath played havoc with this promising beginning. In many parts of the Confederacy the schools were kept going during the first two or three years of fighting, but the collapse of Confederate and state finances closed nearly all of them before the end of 1864. When peace came an effort was made to revive them, but the inability of the people to pay even moderate taxes forced delay. Later, the carpetbag governments undertook to reestablish public schools, but their ill-advised plan of mixing whites and blacks in the same school room closed them effectively to white children. This restored the private school to favor and raised a prejudice against the public school which lasted after its cause had been removed. When reconstruction was over, the schools were set up again with the whites and blacks separated; but they developed slowly in the poorer rural communities because they were largely dependent upon local taxes. In most of the country districts they could be kept open but a few weeks in the year. White prejudice against Negro schools, sharpened by the fact that they must be maintained by the white tax payers, retarded still longer anything approaching adequate provision for the children of the blacks.

The colleges suffered severely in the general collapse.

The few endowed institutions, whose small resources had usually been invested in southern railway stocks or Confederate bonds, were impoverished; the denominational colleges suffered because of the poverty of the church members. The state institutions were in even worse plight, for many of them had been ravaged by federal soldiers and the state governments were unable to repair the damage. Some were virtually closed after the carpetbag governments opened them to Negro students. When the state governments had been restored to white control the state colleges and universities reopened, only to struggle on with pitifully inadequate support from public taxation, hence with poor library and laboratory equipment and with miserably underpaid, if devoted instructors. The effect of this situation upon higher education in the South can hardly be estimated, for before the war there was no great difference in the quality of southern and northern colleges, and in 1860 there had been a higher percentage of southern than of northern youth in institutions of college grade. Following the war the new wealth which poured into the northern institutions and the poverty which retarded the southern brought about the painful contrasts which exist to this day,[190] notwithstanding the marked improvement in southern colleges during the last two decades.

II.

WHEN THE opening of the twentieth century brought an end to the long depression into which the whole civilized world had sunk during the eighteen

---

[190] The 1930s.

nineties, the agricultural South shared in the upward turn. Better prices for crops, especially for cotton, enabled thousands of those farmers who tilled their own lands to get out of debt. Among this class farming methods improved. The efforts of the agricultural colleges, the farm journals and the expert demonstrators of the new experiment stations began to show results in greater diversification of crops, more interest in the conservation of the soil and more attention to improved breeds of farm stock. New labor-saving machinery enabled the farmer to cultivate more acres and thus to increase his output, while the advent of the automobile shortened the road to market and often enabled his children to attend in some neighboring town a better school than the rural community afforded. But his savings were never large, for as he improved his previously low standards of living his small profits were consumed by the increased cost of everything he had to buy. Paradoxically, the rise in the value of farm products was reflected more in the greater prosperity of the merchant and professional classes of the towns than in that of the farmer himself. And notwithstanding the evident improvement in his condition, the home-owning farmer was being slowly but steadily replaced by the tenant. More and more farm property was passing into the hands of the merchants and professional men of the towns.

In 1900 about forty-five per cent of the farms in the cotton states were worked by tenants; by 1910 the percentage had increased to fifty. The tenants shared little in the benefits of the agricultural revival. For the most part untrained in the better methods of farming, moving frequently from one place to another, intent only upon the existing crop, they had no interest in the conservation of the soil, which deteriorated rapidly under their cultivation. Without capital with which to finance the growing of a crop, they still had to be "furnished" by the merchant at exorbitant credit prices, or by the landlord, mortgaging their cotton before it was planted. Diversification was impossible without capital for purchasing tool and livestock and without the consent of both landlord and merchant, who usually insisted upon a large acreage in cotton, the cash crop. The standard of living of both black and white tenants was pitiably low. It is not surprising that the more energetic and ambitious children of the independent farmer as well as of the tenant sought escape from the farm to the more attractive life of the towns. Constantly drained of its best element, with tenantry steadily increasing—by 1920 it had risen to fifty-five per cent—the farm population of the cotton states settled dangerously toward the condition of the poorer sort of European peasantry.

Still bound to a colonial economy and a strangling credit system, more than half of the tillers of the soil in these states remain balanced precariously on the edge of bare subsistence. In comparison with the town dwellers, most of the white farmers are in a far worse position than their great-grandfathers were in 1860. This agricultural section is certainly not the new South, rising unshackled by the destruction of slavery, its energies released, regenerated and prosperous, which Henry Grady visioned some fifty years ago. But however depressed the condition of southern agriculture, there *is* a new South that fulfills in large part Grady's eloquent prophecies. It is the urban South, the product of new industries and commerce.

Perhaps the first authentic sign of the economic recovery of the South after Appomattox was the establishment of new industries and the gradual restoration of others which had collapsed with the Confederacy. Out of the ruck of small log-house tobacco manufacturers in the piedmont of North Carolina emerged a few men of foresight and daring, such as James B. Duke and R. J. Reynolds, who began new ventures and consolidations that grew in to giant corporations. In the hill country of northern Alabama and adjacent parts of Georgia and Tennessee small coal and iron mines and iron furnaces had been established before the Civil War. The richest deposits, in northern Alabama, were long inaccessible ; but in the early seventies a few enthusiasts began a fight for a railroad to the "Red Mountain." When it had been built, Birmingham arose to become the industrial capital of an enormous coal, iron, and steel district. Consolidation followed consolidation until, in 1907, control of the major part of the coal and steel industry passed into the hands of the United States Steel Corporation. In the mean time the great coal and iron deposits of West Virginia, an extension of the Pennsylvania-Ohio region, were being exploited. Throughout the southern Appalachians the empire of iron and steel extended its dominion. Along the piedmont of the Carolinas and Georgia small cotton mills arose to replace those which had been destroyed or left bankrupt at the close of the war. For many years they developed slowly, for their promoters had to rely upon local capital and to fight the jealous and powerful New England interests; but with the advantage of cheap labor and new machinery they gradually took over the major production of all except the finer grades of cloth. The spirit of William Gregg must have rejoiced when the supremacy finally passed from New England to the southern mill region.

Meanwhile the dismantled southern railroads were slowly rebuilt, usually by means of northern credit and often at the price of northern control. The lines in the border states, especially in Kentucky and Tennessee, which had suffered least or which had been repaired for military use by the Union armies, pushed down into the cotton country or acquired bankrupt roads for connection with the Gulf ports. The Louisville and Nashville and the Illinois Central extended to New Orleans. By 1883 new lines stretched from St. Louis into Texas, to Houston, Galveston, and San Antonio. In the same year New Orleans was connected with San Francisco by the Southern Pacific. J. P. Morgan took over for reorganization a group of Virginia roads which in 1894 became the Southern Railway. Through these extensions and combinations the rising factory towns of the Carolina piedmont acquired direct connection with Baltimore, Philadelphia, and New York, while the lower Mississippi Valley and central Texas were suddenly faced about toward Cincinnati, Chicago, and St. Louis.

While southern agriculture staggered under its burdens, a new urban South was rising to prosperity. The railways had set the currents of trade flowing in new directions and in ever greater volume. Inland towns situated at strategic points, like Atlanta, Memphis and Dallas, obtained advantages in competitive rates and became distributing centers for large territories. To them came new business enterprises—railway shops, machine shops, wholesale houses, department stores, banks, hotels, insurance firms, scores of small shop keepers and retail merchant. The growing population, recruited both from the North and the countryside, was slowly acquiring a cosmopolitan character which was in time to modify profoundly southern traditions and social life.

As compared with the growth of the great northern cities that of the southern towns was not impressive, but it was full of significance for the future of the South. Atlanta's population increased from 37,000 in 1880 to 90,000 in 1900; Memphis rose in the same period from 33,000 to 102,000; Dallas from 10,000 to 42,000. The seaports showed a smaller relative growth than the railroad centers of the interior, for in the same twenty years the population of New Orleans rose from 216,000 to 287,000, while Mobile's grew from 29,000 to only 38,000 and Charleston's from 50,000 to a bare 56,000. During the succeeding twenty years, a more prosperous period for business, the southern cities grew apace. Richmond doubled its size, rising from 85,000 in 1900 to 171,000 in 1920; Atlanta rose from 90,000 to 200,000; Birmingham shot up from 38,000 to 178,000; New Orleans from 287,000 to 387,000; Dallas from 42,000 to 159,000; Houston grew from 44,000 to 138,000, and San Antonio from 53,000 to 161,000. These figures arc only for the incorporated areas; in most instances separately incorporated suburbs swelled the true total by many thousands more. Automobiles and improved highways greatly extended the retail trade of the larger towns at the expense of the smaller neighboring villages; they created a vast new business, stimulated travel and quickened the tempo of life everywhere.

The increase of wealth, as of population, was chiefly

in the cities. As taxable values mounted, paved streets, playgrounds and parks multiplied. The urban school systems were enlarged until in buildings, equipment, and curricula they were on a par with those of northern cities. The tax returns from city property enabled the state legislatures, still controlled by the rural districts, to spread a portion of the tax revenues over the rural schools. Automobiles and good roads made it possible for many small country schools to consolidate, build better school houses and employ better teachers. Unfortunately, this development was very uneven, for in many states the rural schools were still shabbily housed, ran for short terms and paid low salaries to teachers.

The course of politics reflected the confusion of a transition period. After the Populist uprising of the nineties had subsided, the Democratic organizations resumed their accustomed control of state affairs; but the Democrats seldom had a definite party policy with respect to local matters and, in the absence of a strong opposing party, fell into divisions between rival aspirants for leadership and power. Nevertheless, a considerable public opinion was forming in behalf of more activity by the state governments in promoting the general social welfare. Prior to 1900 the poverty of the people had kept state revenues and expenditures low and had restricted the functions of the governments to the barest essentials; but now improved economic conditions strengthened the demand of the more progressive element for increased expenditure for public improvements, for public education, and for new agencies for social betterment. This program found its chief obstacle in the unreformed tax systems which still laid the heaviest burden upon real estate. The landowners opposed any increase in this tax, while the business corporations resisted every effort to shift a proportionate share of the burden to other forms of wealth. Likewise, a growing demand for the closer regulation of corporate enterprise, especially with respect to labor policies, met the resistance not only of the corporations affected but of other persons who thought that the development of the South depended upon concessions to capitalistic enterprise and of those conservatives who dreaded to see government interfere with the freedom of business. Improvement in these matters, therefore, went forward unevenly, in some states very slowly. The growth of corporate wealth and the opening of new business opportunities lured the ablest men away from public service, so that the average of political leadership was mediocre. In some states it sank even lower, for this was the period which produced Vardaman, Bilbo, and Blease. Outside the mountain sections, where the Republicans had been strong ever since the close of the Civil War, the Democratic party was apparently as strong as ever; but as big business became more and more powerful and drew more interests to its support, the protective tariff found new adherents and thousands of southern business men ventured openly to express admiration for the attitude of the Republican party toward corporate interests. While they might continue to vote with the Democrats, they prayed for Republican success in the national elections.

The South was being swept into national currents. The war with Spain carried thousands of the sons of Confederate veterans into the army and aroused an enthusiasm for "the flag" which caused the United States

Congress, dominated by Republicans, to remove the last political restrictions upon ex-Confederates. The growth of cities and the shifting of residence, so characteristic of American business life, tended strongly to break down provincial prejudices. The election of Woodrow Wilson to the presidency brought southern political leaders again to the front in national affairs and made the South, for the first time in over fifty years, feel that it was an integral part of the nation. The World War wiped out most of the bitter memories of the Civil War. To hundreds of thousands of school children today that war is as remote as the Revolution against Great Britain.

In similar fashion, for an increasing number of southern people, especially for the younger generation, the traditional culture patterns were breaking down. The greater ease of travel and the constant shifting of business and professional people between northern and southern cities tended to weaken old associations and to set up new standards and attitudes. Thousands of young men and women of the well-to-do urban classes attended northern schools and colleges and made new contacts and friendships. The beginning renascence of southern colleges brought in as instructors scores of young scholars trained in the great graduate schools of northern universities. Some of them, at least, managed to teach the value of scientific methods of investigating not only the physical and biological world, but also the prevailing economic order, the contemporary social organization, and even religious conceptions. Though not always perfectly free to pursue their inquiries far into these mysteries, they called attention to the need for a reexamination of old social values and raised questions which challenged the entire range of traditional concepts and practices. Through books, essays, and uplift magazines the world-wide ferment of discussion spread into the South.

As has always happened, the new teachings stirred up reaction. The great reservoir of conservative tradition was the rural population and the small towns; but a considerable proportion of the urban communities was of rural origin and strongly attached to the same social and religious concepts. Many sincere but literal-minded persons became greatly concerned over the implications which the theory of organic evolution and the higher criticism of the Bible seemed to hold for their religious doctrines, while others convinced themselves that American social life in general was endangered by the introduction of alien ideals. Hence arose that strange phenomenon, the fundamentalist movement, which was not confined to the attacks of religious bigots upon the findings of science but swept the very dregs of provincial prejudice with the organization and activities of the new Ku Klux Klan. It must be remembered, however, that the South had no monopoly on these exhibitions of a belated medievalism.

The militant southern advocacy of prohibition, though sustained largely by the same groups, had a much wider popular support. It was based upon a mixture of motives: a sincere desire to relieve the families of the poor from the hardships entailed by drunken husbands and fathers; the hope of removing temptation from the path of weak men; a determination to destroy the saloon as a corrupting political force; the expectation that the money which went to saloons would be diverted into other channels of trade; and, in the black belt, the necessity of eliminating the danger of drink-crazed Negroes. The evangelical churches furnished the chief impetus to the crusade, which in its later phases acquired a considerable tincture of religious fanaticism. It is one of the historical ironies that the dry South and West pursued exactly the same political tactics in imposing national prohibition upon the wet North as had been used by the North and West in forcing Negro suffrage upon the South—and apparently with the same results.

Although new forces are sweeping through the South with the expansion of industries and commerce and the growth of cities, they have as yet touched lightly the rural population. Tenantry, cotton, and the credit system still hold in thralldom the greater portion of the farming class from the Carolinas to Texas. But as year by year the cities grow while the rural communities remain stationary, the center of gravity shifts steadily toward that urban civilization which seems destined to dominate and standardize American life everywhere. Slowly perhaps, but inevitably, as economic and social changes press them into new paths, the southern people will give up those peculiar attitudes which have grown out of their social inheritance and political experience. Whatever the future of American life, they will share it.

Part Three

More Treatises

Ramsdell's In-Depth Analyses
of Challenges Behind Confederate Lines
that Greatly Affected the War

General Robert E. Lee's Horse Supply, 1862 - 1865[191]

by

Charles W. Ramsdell

With his flank turned and his remaining communications about to be cut, Lee began at once the withdrawal which he had long foreseen must be made. It would have been a difficult operation with his animals in good condition; but now at the end of a severe winter when they were weak and slow from exposure and starvation it was a desperate undertaking. Only the stronger teams were able to take out wagon trains and guns, and on the forced marches without food they soon broke down. The cavalry could not keep pace with the better horses of Sheridan. At the end of a week what was left of a proud army was surrounded and the long struggle was over.

Although it has long been an axiom that the effectiveness of an army depends upon its mobility and its food supply as well as upon its discipline, technical training, and the

[191] Charles W. Ramsdell, "General Robert E. Lee's Horse Supply, 1862-1865", *The American Historical Review*, Vol. 35, No. 4 (Jul., 1930), pp. 758-777. Article and citation are verbatim.

skill of its officers, it is a strange fact that the numerous histories of the military operations of General Lee have paid but little attention to his constant worries over food supply and practically none at all to his equally difficult and equally important problem of maintaining. mobility. These two subjects are so closely interrelated that it is impossible to separate them; but the present study, for the sake of brevity, will deal primarily with one factor in the problem of movement, namely, Lee's supply of horses and mules and his facilities for keeping them in condition for service. It is hoped that the examination of this subject will throw some new light upon Lee's operations, especially during the last two years of the campaigns in Virginia.

The census of 1860 indicates that there was a normal supply of horses and mules in the Confederate States at the beginning of the war; but the lower South was not a horse-breeding country, although a few fine horses for saddle or racing purposes were raised there. The·great horse-breeding region was in Kentucky, Tennessee, and western Virginia, and the planters generally bought their horses and mules from those states. In the principal cotton states, Georgia, Alabama, and Mississippi, there were nearly as many mules as horses, while in Louisiana there were more. In Texas there was a large surplus of horses, but they were mostly of the small "mustang" breed and not well adapted either to draft purposes or to cavalry use. Oxen were widely used throughout the South, especially on the small farms.

In order to understand the situation after hostilities began it is necessary to note the method of procuring and subsisting horses for the Confederate armies. Under an act of the Provisional Congress, March 6, 1861, each mounted volunteer was to furnish his own horse and horse equipment, for which he was to receive forty cents a day and pay for the horse if it should be killed in action. This provision, adopted partly in the interest of economy and partly in the belief that the men would furnish better mounts than the government and that they would take better care of their own property, was later the cause of much difficulty in procuring remounts for the cavalry in Virginia. Mounted officers also furnished their own horses. All other army horses, that is, those for artillery and baggage trains, were to be provided by the quartermaster's department. This department, or bureau, had charge of all army transportation and of furnishing grain and "long forage" *(i.e.,* hay and fodder) for all animals, including those of the cavalry. Horses were used, then, for three purposes: for cavalry and mounted officers, for artillery, and for post and field transportation. Mules also were used in the transportation service, but they were not favored for cavalry or artillery.

Because of the long distances from the lower states to the Virginia front, and also because of the overloaded condition of the railroads, the quartermaster general at first preferred to buy the horses for artillery and field transportation service in Virginia either in that state or in North Carolina. Most of them were obtained in the Shenandoah Valley or in southwestern Virginia. Cavalry regiments that were organized and mounted in other states were generally marched to Virginia. There was no marked difficulty in procuring horses and mules in 1861, but by the summer of 1862 the situation had changed.

The loss of Missouri, Kentucky, western and middle Tennessee, and trans-Allegheny Virginia had cut off the great reservoir of the better grade of horses, while the depreciation of Confederate currency in the spring, after the retreat of Albert Sidney Johnston's army into Mississippi, had sent prices up to unprecedented heights. Early in June the quartermaster general, whose department evidently was not yet well organized, complained that it taxed his exertions and the resources of the country to provide horses for both the transportation and the artillery and to obtain forage for them, and suggested that the number of artillery companies should not be increased.[192]

That astonishing series of operations during the summer of 1862, by which Lee threw the Union armies out of Virginia, had the effect of saving the grain crop to the Confederates. The army became less dependent upon the railroads for food and forage and, except for short intervals, was well supplied until winter came on; but because of the constant service and extraordinary exertions required of them the horses were badly worn down by the end of September, when their food was becoming scarce. The attentive care which Lee always gave to the horses of his army is noticeable throughout this first of his great campaigns.[193] Losses of artillery and

---

[192] Quartermaster General's Letter-Book, I. 362, *et passim,* Confederate Archives, Adjutant General's Office, U. S. War Department. A. C. Myers to G. W. Randolph, secretary of war, June 5, 1862, *ibid.,* IV. 267.

[193] R. E. Lee to Jefferson Davis, Sept. 28, Oct. 1, 1862, U. S. War Department, *The War of the Rebellion: Official Records of the Union and Confederate Armies,* ser. I. vol. XIX., pt. 2, pp. 633, 642-643.

cavalry horses, necessary much exposed in battle, had been severe. With great difficulty the artillery was refurnished with horses; but the problem of remounting the cavalry was harder. Under the law the men must furnish their own mounts. They were to be paid only for horses lost in battle and then only at the value appraised when mustered into service. This allowance was now insufficient, because of the depreciation of the currency, to enable them to buy new ones. Even Virginians who were near home had great difficulty in procuring horses because of the high prices. Men from distant states found it almost impossible to remount themselves. To make matters worse, the horses of Stuart's cavalry had contracted diseases known as "sore tongue" and "greased heel" or "soft hoof" which rendered many of them unfit for use.[194] So many men were dismounted that Lee sought to transfer them to the infantry and to replace them with infantrymen who could procure horses. His cavalry was so greatly weakened that he was unable to operate effectively against McClellan's flank when the latter crossed. into Virginia east of the Blue Ridge late in October. The only means which the secretary of war could suggest for strengthening the cavalry was to purchase one thousand horses in Texas, bring them all the way to Virginia and sell them to the dismounted men at cost.[195]

---

[194] R. E. Lee to G. W. Randolph, Nov. 10, 1862, *Offic. Rec.,* ser. I. v. XIX., pt. 2, p. 709; Heros von Borcke, *Memoirs of the Confederate War for Independence,* pp. 326, 343-344; J. S. Wise, *The Long Arm of Lee;* I. 348. McClellan's cavalry horses were also attacked by these diseases; see, *e.g.,* John Gibbon, *Personal Recollections of the Civil War,* p. 93.

[195] G. W. Randolph to Lee, Nov. 14, 1862, *Offic. Rec.,* ser. I. v. XIX., pt. 2, p. 716.

This suggestion, however, was not carried out.

As the winter of 1862-1863 came on, the difficulty of getting supplies to the Army of Northern Virginia caused Lee great anxiety. Burnside's thrust at Fredericksburg forced him to concentrate his army and thus to decrease his range for foraging and increase his dependence upon the feeble railroads. The immediate countryside was soon exhausted, and·wagons were sent as far as seventy miles away for forage. But the quartermasters' teams were too weak to haul heavy loads so far over bad roads. Sometimes heavy rains or extremely cold weather stopped them altogether. Long distances, light loads, and a growing scarcity of teams and wagons kept the stock of supplies at a dangerously low·level. The railroads were doing but little better, for they were delivering only about one-sixth of the ration of hay and not even that much with regularity.[196] By February food was so scarce for both men and animals that Lee was compelled to scatter his army in order to feed it. Artillery horses were sent as far away as was safe—some towards the James and others to the lower Rappahannock; part of the cavalry was moved to Page County in the Shenandoah Valley; and, most important of all, about half of Longstreet's corps was sent south of the lower James for the double purpose of checking the Federals at Norfolk and Suffolk and of getting out supplies. Lee took the chance that Burnside

---

[196] R. E. Lee to T. J. Jackson, Feb. 7, 1863, *Offic. Rec.,* ser. I. v. LI., pt. 2, pp. 678-679. "Send Hay as fast as possible. . . . The animals here are dying for want of forage and none can be obtained in the country around about." W. H. Kirker, Milford Depot, to G. G. Thompson, Hanover Junction, Feb. 27, 1863, Papers of George G. Thompson (in Library of Congress). Kirker was assistant quartermaster.

would remain inactive, for the bad weather that cut down his own supplies likewise made it difficult for his antagonist to move. There is evidence that, but for the condition of the roads and streams and the lack of food, Lee. himself would have assumed the offensive; for he wrote Jefferson Davis in the middle of February expressing regret that the situation did not admit of attack. The rivers and streams were swollen and without bridges; the roads were impassable; and, he added: "Our horses and mules are in that reduced state that the labor and exposure incident to an attack would result in their destruction, and leave us destitute of the means of transportation."[197]

With the approach of spring and the renewal of active operations Lee gave close attention to his horses. Despite the greatest care, many of them had died during the winter. The quartermaster general was called upon for horses to fill the gaps in the artillery, and every effort was made to help the dismounted cavalrymen to procure mounts. But it was almost a hopeless task. In March, because of "the difficulty of procuring animals and forage, and from the increased demand for transportation and subsistence", Lee ordered a reduction in the transportation of the army—that is, of wagons and teams for the several headquarters and for medical, quartermaster, and other services—to the lowest possible limit. The reorganization of the artillery, and especially the introduction of heavier guns, made heavier draft horses necessary and they were very hard to find. When

---

[197] *Offic. Rec.,* ser. I. v. XXV., pt. *2,* pp. 509, 604, *627,* 632; Douglas S. Freeman, *Lee's Confidential Dispatches to Davis,* pp. 71-72.

requested by General Wade Hampton to increase the number of batteries of horse artillery for service with the cavalry, Lee replied that it was impossible on account of the difficulty of procuring horses. While Hooker was preparing in April to take the offensive, Lee, with Longstreet still absent below the James, was forced to remain immobile because of the condition of his horses and the scarcity of food and forage.[198] The weakness of his

---

[198] "General Orders, no. 43", Mar. 21, 1863, *Offic. Rec.,* ser. I. v. XXV., pt. 2. p. 681; W. N. Pendleton to A. H. Cole, Mar. 31, *ibid.,* p. 695; Lee to Hampton, Mar. 31, *ibid.,* p. 694; Lee to Davis, Apr. 16, *ibid.,* p. 725. In early April, 1863, when the stock of food in Richmond was nearly exhausted, the quartermaster general was unable to supply wagons and teams with which to haul 67,000 bushels of wheat from Essex County on the lower Rappahannock. L. B. Northrop to A. C. Myers, and Myers to Northrop, Apr. 3, 4, 8, 16, 1863, Quartm. Gen.'s "Letters Received", Confederate Archives.

On April 17, 1863, Edmund Ruffin noted in his diary that good hay cost $25 per 100 pounds in Richmond, and added: "It seems to me that our country & cause are now, for the first time during the war, in great peril of defeat—& not from the enemy's arms, but from the scarcity & high prices of provisions, & the impossibility of the government feeding the horses of the army, which is even much more difficult than to feed & support the men. In the cavalry brigade to which my grandson belongs, the horses have rarely had any feed but corn for some months—& are generally without any hay or other long provender, & for weeks together. Horses cannot live on grain alone, even if plentifully supplied with it. As might be expected, the horses are reduced very low in flesh & strength, & many are dying, & more failing entirely. I do not know, but infer that this brigade is not worse supplied than all others of our cavalry in eastern Virginia. And if so, the cavalry and the wagon & artillery teams cannot be capable of performing hard or even moderate service. . . ." *Diary of Edmund Ruffin,* IX. 1623. MSS. Div., Library of Congress. For this and later

cavalry was especially trying, for the Federal cavalry had been heavily reenforced and was beginning to ride over the smaller Confederate units by sheer weight of numbers. This was ominous; for it should be remembered that hitherto the cavalry of Stuart had been superior in fighting power to that of the Federals. The better remount facilities of the Northern army were beginning to tell, and had Hooker known how to use his cavalry at this juncture he might have inflicted disaster upon Lee. On April 20 Lee again ordered a reduction in the scanty transportation of his army. A few days later he wrote to his chief of artillery, General Pendleton, "The destruction of horses in the army is so great that I fear it will be impossible to supply our wants. There are not enough in the country".[199]

General Lee's statement that there were not enough horses in the country for army use after only one year of active warfare may seem surprising; but as he was never given to exaggeration the statement deserves consideration. By "the country" he may have meant Virginia or the region from which his own army normally drew its animals; and he evidently had in mind also the surplus of horses above the minimum requirements of the farmers.[200] Lee, who seemed to consider everything,

---

citations to this diary I am under obligations to Professor Avery O. Craven, who called my attention to them.

[199] *Offic. Rec.* ser. I. v. XXV., pt. *2,* pp. 739, 740-741, 749.

[200] "The waste & destruction of horses in our cavalry service are enormous-& enough to destroy the efficiency of that branch of the army, as well as to increase both public and private expenses beyond all calculation for new supplies of horses. It was one part of our general bad system of southern economy to raise very few horses, & to buy nearly all, & all our

insisted at all times that the farmers be hindered as little as possible in the production of crops because, if for no other reason, his army must be fed largely from Virginia. He had already seen that, because of their weak condition, he could not depend upon the railroads for sufficient supplies from the lower South;[201] besides, the other armies must now draw upon that region. Without teams the Virginia farmers could neither plant nor harvest. It had not taken long to draw off their small surplus of horses and mules; and future levies upon their teams—inevitable, because of the heavy destruction of the animals with the army—must result in a decrease in the production of the food and forage by means of which the army existed and moved. Horses were still to be procured in Virginia, at a heavy cost to agriculture; but Lee was never again adequately supplied with them.

After Chancellorsville, Lee was forced to remain

---

mules, from the western states. A change of this general system of buying to rearing animals, cannot be changed, even under favorable circumstances for obtaining breeders, &c. in less than three year—& the war, & the dangers of every farmer's stock made the circumstances very unfavorable for a change. Therefore there has been little increase in the breeding of horses & mules—the supply by purchases from abroad totally cut off—the waste, by want of food & great hardships & abuse, in our army, made us destructive—as can be conceived—& the raids & robberies of the enemy, in addition, have stripped much of the country of the before diminished & insufficient stock of horses & mules for agricultural labors. This alone is a very serious subject for gloomy anticipations . . . ." *Diary of Edmund Ruffin,* X. 1817-1818 (August 13, 1863).

[201] For a brief account of the condition of the railroads in the Confederacy see the *Am. Hist. Rev.,* XXII. 794-810.

immobile for more than a month, partly because he needed time for the reorganization of his army after the death of Stonewall Jackson, but partly also because of the condition of his horses.[202] The artillery horses were again sent away to be foraged; heavy horses were transferred from the transport service to the guns and replaced by mules; new ones were brought up by the quartermasters. The cavalry was still in bad condition. More than one-fourth of Stuart's men were without mounts, while nearly all the horses in service were poor and weak. The cavalry was given as much rest as the safety of the army would allow, and as spring grass and clover came on the animals began to mend. Lee was greatly cheered in the last days of May by the capture of some 1200 to 1500 horses during a raid on the upper Potomac by Generals W. E. Jones and Imboden; but even with this addition to his resources he was unable to provide sufficient teams for the medical wagons, ambulances, and ammunition trains, and was apprehensive that he must reduce the artillery.[203] But he proceeded, nevertheless, to manoeuver Hooker out of Virginia into Maryland, and to take his own army into Pennsylvania.

As Lee moved north towards Gettysburg, the Federals endeavored to remove all horses from his reach; but he obtained some, though evidently not enough to replace those killed or worn out and left behind. When he was back in Virginia in the latter part of July, straining every resource to recover from that disastrous expedition,

---

[202] For a vivid description of the sufferings and losses of cavalry horses in both Federal and Confederate service see a letter from Charles Francis Adams, jr., to his mother, May 12, 1863, in *A Cycle of Adams Letters,* II., pp. 3-5.

[203] *Offic. Rec.,* ser. I. v. XXV., pt. 2, pp. 808, 809, 820, 825.

he again gave anxious consideration to the condition of his horses. All the animals were greatly weakened by the strain of the campaign. As the corn crop was not yet matured, grain was very difficult to get in northern Virginia; and the railroads could not deliver enough from the south. Fortunately, Meade crossed slowly into Virginia and kept out of striking distance east of the Blue Ridge, and Lee used the respite allowed him to send away to refreshment camps the horses that were utterly broken down.[204] But he was again reduced, practically, to immobility. On August 24, he wrote President Davis:

> Nothing prevents my advancing now [against Meade] but the fear of killing our artillery horses. They are much reduced, and the hot weather and scarce forage keeps them so. The cavalry also suffer and I fear to set them at work. Some days we get a pound of corn per horse and some days more; some none. Our limit is five per day per horse. You can judge of our prospects . . . . Everything is being done by me that can be to recruit the horses. I have been obliged to diminish the number of guns in the artillery, and fear I shall have to lose more.[205]

---

[204] Lee to J. A. Seddon, secretary of war, Aug. 7, 1863, *Offic. Rec.,* ser. I. v. XXIX., pt. 2, p. 628. Lee to Longstreet, July 19, 1863, and "General Orders, no. 25" [Stuart's], July *29,* 1863, *ibid.,* ser. I. v. XXVII., pt. 3, pp. 1o24, 1050.

[205] *Offic. Rec.,* ser. I. v. XXIX., pt. 2, pp. 664-665. The Federal ration of grain to horses was ten pounds per day, though it probably averaged about eight. *Cycle of Adqms Letters,* II. 3.

In October, when the local corn crop was available and his horses were in somewhat better condition, Lee moved against Meade's right and forced him back to the line of Bull Run; but as the country thereabouts was bare of supplies he returned to the Rapidan.

As the difficulty of procuring fresh horses increased, greater attention was given to the care of those disabled. Hitherto, it seems, these animals had been turned over at stated intervals to quarter-master officers or agents, who distributed them on pastures under the care of subordinates. Here they received little attention and were left to recover or die. They were seldom properly inspected by veterinaries, and diseased horses were often placed with the others with the result that the disease spread. The new plan adopted in the fall of 1863 seems to have originated with General W. N. Pendleton, Lee's chief of artillery. As the plan was worked out, the whole Confederacy was divided into four inspection districts for field transportation, under an inspector general at Richmond, Major A. H. Cole, who was on the staff of the quartermaster general. The First District, comprising Virginia and North Carolina, was placed under Major George Johnston.[206] The distinctive feature of the plan was the establishment of "horse infirmaries" under special officers. The infirmary headquarters for Virginia were placed at Lynchburg, under Major J. G. Paxton. As unserviceable horses and mules were withdrawn from the army, they were to be examined carefully, the diseased

---

[206] Pendleton to Lee, Aug. 13, 1863, and Pendleton to A. H. Cole, Sept. 3, 1863, *Offic. Rec.*, ser. I. v. XXIX., pt. 2, pp. 643, *697.* Order Book, Inspector of Field Transportation, Oct. 7, 1863, *et passim,* Confederate Archives.

ones segregated, and the rest distributed under experienced caretakers in those counties about Lynchburg remote from army operations where feed and pasturage were most plentiful. The results of this system will be described later.

As the winter of 1863-1864 drew on, the perennial spectre of famine threatening men and animals again haunted Lee. Even with the army reduced by the absence of Longstreet's corps in Tennessee, the railroads were not bringing enough food. Moreover, large portions of the upper Virginia piedmont and the lower Valley were almost bare of grain and forage. The horses, overworked and underfed, were so poor they could hardly be used. In November, Meade threatened an advance across the Rappahannock—the route Grant took in May the next year—and Lee wrote Davis:

> Should he move in that direction, I will endeavor to follow him and bring him to battle, but I do not see how I can do it without the greatest difficulty. The country through which he [we?] will have to pass is barren. We have no forage on hand and very little prospect of getting any from Richmond. I fear our horses will die in great numbers, and, in fact, I do not know how they will survive two or three days' march without food.[207]

Fortunately, Meade did not push forward in earnest

[207] Lee to Davis. Nov. 12, 1863, *Offic. Rec.,* ser. I. v. XXIX., pt. *2,* p. 832.

and, after the affair of Mine Run, the two armies settled down in winter quarters. In August, 1863, Brigadier General A. R. Lawton, of Georgia, had become chief of the quartermaster's department, and new vigor was soon evident in that important bureau of supplies. Though the railroads could not be strengthened, their transportation service was more diligently supervised and the supply of corn from the Carolinas and Georgia flowed somewhat more steadily to the Virginia depots and camps. But it was never enough. Lee was obliged to monopolize the stocks of corn and forage near the Virginia railroads and the James River Canal; but this did not suffice and by December the horses were failing rapidly. To make the situation worse, the dreaded disease glanders appeared among them. The artillery horses were moved to fresh camps farther west. The cavalry, which must always be on the alert, was in a pitiful state.[208] So many of Wade Hampton's men were dismounted that he begged permission to move two brigades back to South Carolina to enable his men to procure new mounts-a proposal which Lee dared not adopt. When it was proposed that the cavalrymen be mounted on government horses, Lee remarked that he did not see how the horses could be procured, since not enough could be had for the artillery and transportation. An undated memorandum, evidently prepared by some officer in the field transportation service about this time, also opposed the proposition on similar grounds with the additional argument that the

---

[208] *Offic. Rec.,* ser. I. v. XXIX., pt. 2, p. 835. On the difficulty of foraging the cavalry, see H. B. McClellan to Wade Hampton, Nov. 11, 1863, and Thos. L. Rosser to T. G. Barker, Nov. 14, 1863, *ibid.,* ser. I. v. LI., pt. 2, pp. 783, 786.

volunteer could purchase a horse through a friend or neighbor when the government agent could not, and that the soldier would take better care of his own than of a government horse.[209]

The situation grew worse during the winter. On January 3, 1864, the commissary officer for Richmond reported that the entire stock of breadstuffs in that city had been exhausted and that no requisitions for Lee's army could be filled until the railroads from the south could increase their shipments.[210] Evidently the shipments were increased, but food remained scarce both in Richmond and in the camps. Worried over the scarcity of horses, Lee asked for fresh cavalry regiments said to be in South Carolina and Georgia, but did not get them. General J. E. B. Stuart pointed out that only well-to-do men could now buy horses for cavalry service. The heavy losses among cavalry horses are illustrated by a report of General Wade Hampton, on February 1, 1864, that although 2000 horses had been brought on to one of his brigades (Butler's) within the past year, besides many that had been captured, not 500 men could be mounted. In February Federal cavalry was raiding with impunity in

---

[209] Hampton to Lee, Dec. 7, 1863, *Offic. Rec.,* ser. I. v. XXIX., pt. 2, pp. 862-863. Lee to Davis, Nov. 29, *ibid.,* p. 853. Papers of Field Transportation Office, in Confederate Archives.

[210] "The reserve of flour and hard bread has been consumed, and the receipts of corn for the past week have been totally inadequate to our daily wants. The accumulations at Greensborough and Charlotte still remain unmoved, only fifty-four cars having arrived at Danville from Greensborough during a period of four days, while the wants of this Department alone demand the use of eighty cars for the same time." Maj. S. B. French to Col. L. B. Northrop, January 3, 1864. *Offic. Rec.,* ser. I. v. LI., pt. 2, p. 808.

the vicinity of Richmond because the Confederate cavalry, without forage for the jaded horses, was unable to follow the raiders.[211]

Under these conditions Lee had no choice but to remain on the defensive. In January, Longstreet, then near Knoxville, had suggested that Lee move forward in March toward Washington, while he himself should advance into Kentucky. Lee replied:

> ... You know how exhausted the country is between here and the Potomac; there is nothing for man or horse. Everything must be carried. How is that to be done with weak transportation on roads in the condition we may expect in March ? ... After you get into Kentucky I suppose provisions can be obtained. But if saddles, etc. could be procured in time, where can the horses be ? They cannot be obtained in this section of country, and, as far. as my information extends, not in the Confederacy.[212]

Even in the Shenandoah Valley the Confederate forces were unable to occupy the region north of Staunton

---

[211] Lee to Seddon, Jan. 23, 1864, *Offic. Rec.,* ser. I. v. XXXIII. 1118; also, inspection report of S. W. Melton to S. Cooper, Mar. 14, *ibid.,* v. LI., pt. 2, pp. 835-837; Stuart to S. Cooper, Jan. 28, *ibid.,* v. XXXIII. 1126; report of Hampton, *ibid.,* p. 1140; Hampton to Lee, Feb. 12, and Lee to Elzey, Feb. 18, *ibid.,* pp. 1152, 1185.

[212] A. L. Long, *Memoirs of Robert E. Lee,* pp. 637-638.

because both the grain and the long forage there were exhausted. In April, Lee expressed a desire to strike boldly at the enemy on the Rappahannock; but both his artillery and cavalry horses were widely scattered for foraging and he could not bring them to the army because he could not feed them there. As spring opened, the horses were able to get some grass and clover, and they began to improve; but they were far from being in condition for the strenuous campaign that was impending. There was no grain in the country near the Rapidan; the railways could not bring a full supply; nor could the wagon trains haul enough from the depots to enable Lee to concentrate against the thrust which he knew Grant was preparing. At this critical time the ordnance bureau proposed to solve Lee's difficulty about artillery horses by taking away some of his artillery.[213]

When Grant crossed the Rapidan with his well-equipped army early in May, Lee gathered his forces as rapidly as possible and struck the Federal advance in the Wilderness. Until the last minute possible he had been obliged to keep his army scattered in order to feed it. In the campaigns which followed, around to the Richmond and Petersburg fronts, the loss of horses both from the casualties of battle and from overwork was extraordinarily heavy. By the end of May many of the field batteries were practically out of service through lack of horses.[214] Although some fresh animals were obtained,

---

[213] J. D. Imboden to J. A. Early, Feb. *22,* 1864, Lee to Davis, April 15, and Lee to Bragg, April 16, 1864, *Offic. Rec.,* ser. I. v. XXXIII. 1194-1195, 1282-1283, 1285.

[214] Reports of John Esten Cooke to Pendleton, May *25* and 29, 1864, *Offic. Rec.,* ser. I. v. XXXVI., pt. 3, pp. 830, 847. P. W. Alexander, army correspondent of the Savannah

the quartermaster general sent back a requisition for artillery horses with the suggestion "that the proportion of-field artillery be reduced" because of "the great scarcity of animals throughout the Confederacy and the necessity of having enough for agricultural purposes".[215] But the cavalry, still dependent for remounts upon personal purchases by the men, was in worse plight. Not only had the losses been heavy and the replacements scanty, but the horses remaining were so badly broken down by hard riding and lack of forage that they were incapable of long marches.[216] Grant's cavalry, now under Sheridan, was

Republican and an unusually frank and reliable writer, wrote his paper on July 29, 1864: "Just after the battle of the Wilderness the railway lines in Lee's rear were cut by the enemy, and his animals reduced almost to starvation. The county of Spottsylvania is poor, and there was no grass for the horses, which suffered so much that it was with great difficulty they could draw the wagons and artillery when the army moved. Under these threatening circumstances, the people in that and the neighboring counties, who had already paid their tithes and been stripped of nearly everything they had, were applied to by the authorities to *loan* the Government all the corn and other supplies they could spare. The response was unanimous: The heroic men and women said Lee's brave army should have all that was necessary to carry them over the difficulty, even if they had to go without themselves. . . . Thus Lee's hands were held up until the great battles of Spottsylvania were fought. . . ." Savannah *Republican,* August 5, 1864. It is unnecessary to point out the hazards of an army which in such circumstances must rely upon such expedients.

[215] Endorsement of A. R. Lawton, June 28, 1864, Quartm. Gen.'s Office, "Letters Received and endorsements", v. XI., no. 158, Confederate Archives.

[216] On May *27,* after noting down accounts of the burning of houses and barns, destruction of food supplies, and the stealing or killing of animals, etc., by Federal raiders, Edmund Ruffin penned this reflection: "If this policy of the enemy is

active and aggressive as well as numerically superior. Wade Hampton, worthy successor to the lamented "Jeb" Stuart, was taxed to the uttermost to ward off Sheridan's thrusts at Lee's lines of communications.

Although Lee had gained some advantages by the removal to the new front on the Richmond-Petersburg line, they did not fully compensate him for the new difficulties which he now had to face. He had shortened his railway connection with the Carolinas and Georgia, whence most of his supplies must come; and the region immediately south and west of Richmond was not so completely denuded of food-stuffs as was that near the Rapidan. But because of the long line of works he was obliged to defend with his relatively small army, he could not attack Grant and was deprived of any favorable opportunity for offensive manoeuver. He did not fear a frontal attack; but he was fully aware of the danger to his

---

permitted to proceed . . . & Grant will hold off from giving battle to any army opposed to him, the result must be the reduction of Richmond & of Va., not by arms, but by starvation of the country & destitution of our armies. . . . Our cavalry is almost worthless for fighting, because of the broken-down condition of a large proportion of the horses, & the inability to replace them from any surplus stock of the country, & the impossibility of providing half enough provender. Yet, the Yankees, by plundering, take every serviceable horse left for agricultural & private uses, & provide themselves abundantly with forage, even from the most destitute localities. . . ." *Diary of Edmund Ruffin,* XII. 2124. A month later he recorded the failure of Confederate cavalry to pursue some Federal raiders and commented: "This is regularly the case in all raids of the enemy. They, by stealing fresh horses, & other facilities, are able to keep better mounted, & their horses better fed, & cannot often be overtaken, or matched in battle, by our cavalry, on half-starved & failing horses. . . ." *Ibid.,* p. 2178.

railway communications on his flanks. He must depend upon his cavalry to guard, on the one side, the Virginia Central Railroad, lest his communications with the Shenandoah Valley be broken, and on the other, the Petersburg and Weldon road, which connected him with the seaboard of the Carolinas. This last was the most efficient of all the roads which carried supplies to his army, but it was also the most exposed to attack. The Southside Railroad, running west from Petersburg to Lynchburg, and the Richmond and Danville were better protected, but they were both "neighborhood" railroads and were too weak to stand heavy traffic.[217] The Richmond and Danville had recently been connected with western North Carolina by means of the newly built Piedmont Railroad (Greensboro to Danville); but so frail was the Piedmont, a badly constructed narrow gauge, that it could bring to Danville only small shipments, which had to be reloaded for forwarding to Richmond. Every heavy rain washed out track or bridges, and it was necessary to put on wagon-trains between Greensboro and Danville to supplement the efforts of the little railroad. If Grant should break his southern Communications, Lee would be forced out of Richmond and must move back toward the piedmont region of North Carolina. But Richmond was the only railroad center of strategic importance north of Columbia, South Carolina, and the loss of its foundries, shops, factories, and supply depots would be disastrous, to say nothing of the effect which the loss of the capital must have upon public morale. Moreover, it was a question whether, with

---

[217] A. R. Lawton to Lee, June 23, 1864, Quartm. Gen.'s Letter Book. VIII. 308.

his transportation facilities so weak, he could withdraw successfully.

Grant, of course, saw the situation just as clearly as Lee. Early in June, even before he had touched the James, he ordered Hunter up from western Virginia to attack Lynchburg and sent Sheridan with a heavy force of cavalry to strike the Virginia Central and cooperate with Hunter. If successful, they would break Lee's western and northwestern communications. Hunter was repulsed; and Sheridan was checked at Trevilian's Station by Wade Hampton. Late in June, with his base firmly established on the James and a secure line of communication by water, Grant sent Wilson on another raid south of Petersburg across the Weldon road and against the Southside and Danville lines. Though he was severely handled, Wilson did some damage. A few days later, July 5, Lee wrote to Jefferson Davis that the numerical superiority of the enemy's cavalry caused him serious apprehensions about the safety of his southern communications, and that if these were lost he "need not point out the consequences". He did not know where any reenforcements could be had, but suggested that horses might be brought from Texas by swimming the Mississippi, and that others might be obtained from within the enemy's lines in western and northwestern Virginia by bartering cotton and tobacco for them. Upon "obtaining an increase of our supply of horses, and recruiting our cavalry . . . I believe, depends the issue of the campaign in Va."[218] Nothing could be more significant than this statement, with the proposal to get the needed

[218] Douglas S. Freeman. *Lee's Confidential Dispatches to Davis,* p. 273.

horses from the enemy's territory and from far-off Texas.

The damages done to the railroads by the Federal cavalry raids were soon repaired, and Grant did not repeat this experiment. His own losses in horses had been severe, and possibly he had found it difficult for his cavalry to operate very far within Confederate territory where grain was scarce. But he continued his attacks on the Weldon railroad and, after several failures, at length got footing on it in August. Lee could not dislodge him; but through the energetic and resourceful efforts of Quartermaster General Lawton wagon-trains were sent around the break and supplies continued to come through from the south. But the wagons could not bring as much as the railroad had brought, and the margin above absolute destitution became perilously small. No surplus of corn could be accumulated for the coming winter. To send by railroad bulky articles like hay or fodder for the horses was out of the question. There was little grass or clover in the wooded area in which the army now lay, and, whenever the situation permitted, both artillery and cavalry horses were sent back to better pasturage.[219]

The most energetic efforts failed to obtain a sufficient supply of fresh horses and mules during the summer. The farmers who had been robbed of their teams during the Federal cavalry raids had barely enough left for harvesting their crops and planting their fall wheat. In those Virginia counties subject to Federal raids many farmers were without any horses at all that were fit to work, and they were now calling upon the government to

[219] Lee to Hampton, July 22, 1864. *Offic. Rec.,* ser. I. v. XXXVII., pt. 2, p. 598; inspection report of H. E. Peyton to S. Cooper, Sept. 23, 1864, *ibid.,* v. XLII., pt. 2, pp. 1270-1278.

supply them with others.[220] The schedule prices for both horses and grain as fixed by the impressment commissioners, for government purchases, in Virginia and North Carolina were so far below market 'prices that farmers were unwilling to part with the few horses or the scanty grain still left them, even for Lee's army. The market value of good horses and mules, fit for army use, in the summer of 1864 seems to have varied from $1500 to $2500 (Confederate currency); the average impressment price was $500, except during the month of July, when it was $1000. General T. H. Holmes, commanding in North Carolina, complained that he was obliged to pay $4000 "for a very common one" for his personal use.[221] Impressment officers were directed to

[220] Wm. A. Staples, a farmer of Bedford County, Virginia, wrote to Secretary Seddon, August 12, 1864: "Hunter's men on their late Raid stayed three days on me taking 3 Negro men all my Horses oxen fat cattle Bacon Corn Flour Sugar Coffee Molasses all Harness Saddles Bridles all our clothes·breaking crockery ware. . . . I own a Farm 450 acres on the R. R. in the finest district of Bedford, work 9 hands-have 100 acres Clover to fallow for Wheat but have only two Horses & they old broken down ones I found left by the enemy. I have no money to buy Teams takes all to pay expenses & for provisions until Fall. No horses are for sale if I had money. I heard that the Gov't turn over Horses to those who lost by the enemy. I have waited until now the last moment. If you can give me an order on Maj. Paxton Q M at Lynchburg (agent for all this section has control of several thousand Govt Horses at pasture) for 3 Horses I can now fallow and seed 100 Bus. Wheat on finest land here if the Horses are good Farm Horses. No others are of any value. It would be to the interest of the Gov't. . . ." When this request was referred to Major Paxton he replied that he had no horses to spare. Quartm. Gen.'s "Letters Received",·Confederate Archives.

[221] Sixteen horses taken by Hunter's men in June from Mrs. M.

make an estimate of all horses and mules on each farm, not absolutely necessary to make the growing crop, and to take the surplus, seizing pleasure horses first.[222]

When the summer of 1864 had drawn to an end, more than one-fourth of the cavalrymen were still dismounted and infantrymen who could procure horses were being transferred to·the cavalry.[223] Many batteries of the field artillery were without any horses at all. These batteries were put into the defenses, but their immobility was a serious matter when a shift was necessary to meet an attack at another point. Lee called the attention of the secretary of war to the scarcity and inferiority of the artillery horses, and asked whether there was any prospect of relief. Seddon answered that the quartermaster general would endeavor to get horses from

---

C. Massie, of Nelson County, were appraised, under oath, by three neighbors at an average valuation of $1450; but several of them were evidently not fit for the army. The highest single valuation was $2500. Massie Papers, in the University of Texas Library. For schedule of prices, see *Offic. Rec.,* ser. I. v. XL., pt. 3, pp. 766-769*;* v. XLII., pt. 2, pp. 1151-1155; also, Lt. Gen. T. H. Holmes to A. H. Cole, July *22,* 1864, Quartm. Gen.'s "Letters Received", in Confederate Archives.

[222] "Y0u will take pains to impress upon the citizens the urgent demand there now exists in Gen. Lee's Army, for horses and mules. There are now many Batteries of Artillery inactive for the want of horses to pull them; and entire Brigades are without the necessary teams to supply them with provisions." Jas.·N. Edmonson, Inspector Field Transportation, Greensboro, N. C., to Capt. C. R. King, June 20, 1864. *The Papers of Thomas Ruffin,* III. 398 (Hamilton ed., Raleigh, 1920). / AM. HIST. REV., VOL. XXXV.—53 [sic]

[223] Abstract from return, cavalry corps, Sept. 30, 1864, *Offic. Rec.,* ser. I. v. XLII., pt. 2, p. 1309; see also, Wade Hampton to Lee, Oct. 24, and Nov. 2, 1864, *ibid.,* pt. 3, pp. 1161-1162, 1198-1199.

within the enemy's lines and mules from Mexico. General Lawton and Colonel A. H. Cole, inspector general of field transportation, did in fact investigate the Mexican market and learned that the prospects for a supply of mules were good, provided that a discreet and reliable agent were sent with gold or sterling exchange.[224] Dilatoriness either in the war department or in the treasury delayed the selection of an agent until February, 1865, when it was too late. Meanwhile there was no relief. The cavalry was so depleted by December that Longstreet advised that the men be mounted on mules. For various reasons few "recruited" animals were being returned from the horse infirmaries.

The horse and mule infirmary established in the Lynchburg region in October, 1863, had not succeeded as well as had been hoped for, but probably as well as was possible under all the circumstances. The difficulty was that no animals were sent to it until exhausted. Many of these had developed glanders and had to be killed; of the others many were too exhausted to recover. A report of the officer in charge, Major J. G. Paxton, on February 6, 1865, states that during the whole period of approximately fifteen months, he had received 6875 horses, of which only 1057 had been recruited and

---

[224] Lee to Seddon, Oct. 4, and Seddon to Lee, Oct. 5, 1864, *Offic. Rec.,* ser. I. v. XLII., pt. 3, pp. 1134, 1135-1136; Levin Lake, Meridian, Miss., to Maj. A. M. Paxton, Oct. 28, 1864, forwarded to Gen. A. R. Lawton, Quartm. Gen.'s "Letters Received", Confederate Archives; also, endorsements of A. R. Lawton, Nov. 11 and 13, 1864, Quartm. Gen.'s "Letters Received and Endorsements", XII., nos. 157, 202, Confederate Archives; and A. H. Cole to Lawton, Nov. 17, 1864, "Letters Received", *ibid.*

returned to the army, 2844 had died, 133 had been lost or stolen, 559 had been condemned and sold, 799 had been transferred to an infirmary in North Carolina, and the rest, 1483, were still unserviceable. Of mules, 2885 had been received, of which 1644 had been recruited and returned, 575 had died, a few had been sold as hopeless, and the rest were still unrecruited.[225] The mules made a much better showing than the horses, for while only 15 per cent. of the horses had been returned to service, 57 per cent. of the mules had recovered. Paxton estimated that the average life of a horse in the artillery and transportation services was seven and a half months, and that a mule was five times more durable than a horse. But the mortality among artillery horses was much greater than in the transportation service and it was heavier still in the cavalry. Paxton claimed that he and his agents had purchased or impressed during the fifteen months 4929 horses and mules in Virginia at an average price of $524.20, and that he had had great difficulty in getting the funds with which to pay for them. His estimate of the number of animals required in Lee's army and neighboring posts for artillery and transportation was 7000 horses and 14,000 mules every fifteen months. His figures can not be verified, but an estimate of the animals in all varieties of transportation service in that army on November 4, 1864, gives 1321 horses and 12,316 mules. This does not include cavalry mounts or artillery horses. By December, 1864, corn and forage for only about 1000 horses could be provided in the region about Lynchburg and to the south of it in Virginia. About 600 were

---

[225] This report is in manuscript among the Personal Papers of Jas. G. Paxton, Confederate Archives.

quartered in northern North Carolina, and it was planned to send 1000 more into the southern part of that state until it was learned that the commissary officers claimed all the surplus grain there.[226] At the date of Paxton's report, February 6, 1865, more than 4000 cavalry horses from Lee's army were in infirmaries in South Carolina, mostly in Lancaster County.

When the winter of 1864 closed down on Lee's army the familiar difficulties of finding food and forage were infinitely worse than ever The Shenandoah Valley, devastated by Sheridan, could furnish nothing, and horses there were dying of starvation by hundreds on the farms. All the country within reach of the army was swept bare of supplies. Since the currency was worthless, the purchasing officers and agents could not buy provisions with it and were forced to resort to barter or impressment; but the country along the railroads had already been combed and the scarcity there was so extreme that not even coin could have procured enough food for the army. Nor could the worn and crippled railroads have brought enough if there had been no scarcity in the Carolinas. The men were on one-fourth rations and some days had none at all. The winter was extremely severe upon both men and animals. The hungry and half-frozen men were deserting in large numbers; but the famished horses could only die unless removed. In January, Lee had to diminish his cavalry still further by sending Butler's division to South Carolina to

---

[226] Summary statement in Personal Papers of Major George Johnston, Confederate Archives. Johnston was chief inspector of field transportation at Richmond. J. G. Paxton to Q. M. G. O., "Letters Received", Confederate Archive. The letter is undated but is found with others of December, 1864.

get fresh horses which were to be collected by the government.[227] General Hampton also went down to superintend recruiting. Neither he nor Butler's division ever returned to Lee, for they were retained to operate against Sherman.[228] Unserviceable cavalry horses to the number of 2700 were sent to the same state at about the same time to be foraged; but they had to be scattered over a wide territory far back from the railroads where the tithe-gathering officers did not operate zealously. Lee had now only two weak divisions of cavalry with his army. In order to procure forage, the greater part of W. H. F. Lee's division had to be sent forty miles away, by the roads, to Stony Creek, beyond the gap in the Weldon railroad on the south. Here the horses were kept in fair condition; but they were too far away to be of use in emergencies, and they could not be fed when brought up to the army. Fitzhugh Lee's cavalry, on the left flank and north of the James, was farther from supplies, and the horses were in such bad condition that they were unfit for hard service. The artillery animals were sent back, only a few being left with the guns. The men of some of the field batteries were sent to the heavy guns in the fortifications because their

---

[227] M. G. Harman to R. M. T. Hunter and A. T. Caperton, Jan. 17, 1865, *Offic. Rec.,* ser. I. v. XLVI., pt. 2, p. 1110. Lee to Seddon, Jan. 11, *ibid.,* p. 1035. Lee to Cooper, Jan. 19, *ibid.,* p. 1100; .also "Special Orders, no. 8" [Hampton's], *ibid.* p. 1101.

[228] General Hampton failed to procure sufficient funds for the horses he needed; and on February *2,* the presidents of eight state banks met at Columbia in the office of Governor Magrath and agreed to advance to the state $1,000,000 for the purchase of horses for Hampton. Printed circular agreement in "South Carolina: Letters Received by the Governor", Confederate Archives.

own batteries could not be supplied with horses.[229] Because of the weakness of the cavalry which guarded his flanks, Lee was obliged to extend his already too thin lines. In the face of an active and aggressive enemy this was dangerous business, but there was nothing else to do. He saw clearly what was in store for him, and repeatedly pointed out that he could not continue to hold Richmond without more men and horses and food. The government, however, was really helpless. The purchasing officers had to contend not only with an actual scarcity of supplies but also with a collapsed currency which paralyzed every effort.

It was now proposed to change the law which required the cavalrymen to furnish their own mounts. One evil in the existing system was that whenever a man was dismounted he had to be furloughed home in order to find another horse. In consequence, many men were long absent from service when needed at the front. In order to get the cherished furloughs some of the homesick men deliberately disabled their own horses. With the approval of the officers most concerned with the problem, a bill was introduced into the Confederate Congress on December 29, 1864, to require the quartermaster general to provide horses for dismounted cavalrymen and to purchase the horses of any cavalry unit upon recommendation of the general commanding in the

---

[229] J. G. Paxton to A. R. Lawton, Jan. 27, 1865, "Personal Papers of Jas. G. Paxton", Confederate Archives. Inspection report of Maj. Geo. Freaner, March 1, 1865, MS. in Confederate Archives. W. N. Pendleton to T. H. Carter, Jan. 17, 1865, *Offic. Rec.,* ser. I. v. XLVI., pt. 2, pp. 1083-1084. Pendleton to W. H. Taylor, March 18, 1865, *ibid.,* pt. 3, pp. 1322-1324.

field.[230]

The bill passed both houses on February 14 and was approved by Jefferson Davis on February 23. It is doubtful whether it was ever put into effect at all; but at that late day it could not have relieved the situation to any appreciable extent. If the prospect for fresh horses had been bad in the early winter, it was desperate by the end of January. General Pendleton, chief of artillery, suggested that the unserviceable horses be turned over to the farmers in return for good horses impressed. This would save the scanty stock of forage in the recruiting depots. "The question of our horse supply", he said, "is hardly second to that of supplying men for the army, or food for the men."[231] Major A. H. Cole, who was charged with the duty of providing horses and mules for artillery and transportation service for all the armies east of the Mississippi, had been making estimates of the number of animals that would be required and canvassing the means of procuring them. In two communications to General Lawton, written the same day, February 1, he reviewed the situation. He estimated that the armies would require for the spring service some 6000 additional horses and 4500 mules. The number to be had by impressment would depend upon how many could be taken safely from agriculture. Evidently Cole thought that no more could be

---

[230] A. H. Cole to A. R. Lawton, December 24, 1864. Quartm. Gen.'s "Letters Received", Confederate Archives. *Journals of the Confederate Congress,* vol. 7 (House Journal), 400, 419, 513, 543-544, 577, 650; vol. 4 (Senate Journal), 498-499, 544. The law was never officially printed and portions only appear in the *Journals.*

[231] W. N. Pendleton to A. H. Cole, Feb. 7, 1865, *Offic. Rec.,* ser. I. v. XLVI., pt. 2, p. 1208.

taken from that source, for, as already stated, he suggested that all should be procured from within the enemy's lines and from Mexico. He thought that 5000 could be got from regions occupied by the Federals east of the Mississippi—3000 from within Mississippi and 2000 from Virginia and North Carolina. In the last two states gold or United States currency would be necessary; and prices would range from $60 in gold for first class, to $40 for second class, animals. In Mississippi cotton must be furnished the purchasing agents at the rate of 600 pounds for first class horses, and the agents must be allowed to work without interference from the treasury officials. Gold or sterling exchange would be essential in Mexico. Some exchange had been furnished in December, but the failure to appoint a suitable agent had made it useless.[232]

On February 14, Cole estimated that the calls for horses and mules for *immediate* service in Virginia and North Carolina alone aggregated 3200 horses and 2400 mules. At the same time he was expected to furnish 2650 animals for the forces gathering in the Carolinas to oppose the northward march of Sherman. There was no time to look beyond the Mississippi. There was no other recourse but to impress from the scanty supply of the farmers, for which $3,000,000 in currency was necessary at once, and to purchase from across the lines, for which $100,000 in gold was essential. The attitude of the farmers as well as that of the state officials made impressment a failure. A week later Cole advised Lee that he was getting no animals whatever for the army for the reason that he had received no gold from the treasury.

---

[232] *Offic. Rec.*, ser. I. v. LXVI., pt. .2, pp. 1190-1191; also, *ibid.*, ser. IV. v. III. 1087-1089.

When General Lee suggested that the government convert its cotton and tobacco into gold for this purpose, the secretary of the treasury insisted that the effort had been made to do so and promised that it would be continued. The government had now, however, been reduced to the slow process of barter and was really unable to act promptly. The only evidence found that any of the gold was ever furnished is an order of Cole, on March 7, turning over $2000 in coin to a bonded agent for the purchase of animals within the enemy's lines.[233]

No evidence has been found that Lee ever received any of the horses he called for in February and March. Again and again he called the attention of the secretary of war to his perilous situation and begged for food and forage for the army. In his famous letter of March 9 in which he reviewed the military situation with vivid frankness, he said :

> Unless the men and animals can be subsisted, the army cannot be kept together, and our present lines must be abandoned. Nor can it be moved to any other position where it can operate to advantage without provisions to enable it to move in a body.[234]

---

[233] A. H. Cole to Lawton, Feb. 14, and Cole to Col. Corley, with endorsements, Feb. 20, 1865, *Offic. Rec.,* ser. I. v. XLVI., pt. 2, pp. 1232-1233, 1242-1243. Order Book of A. H. Cole, Inspector General of Field Transportation, p. 48, Confederate Archives.

[234] Lee to Breckenridge, March 9, 1865, *Offic. Rec.,* ser. I. v. XLVI., pt. 2, p. 1295.

Although he knew that Grant was preparing to turn his right at Hatcher's Run, he still had to keep his little cavalry force on that wing down at Stony Creek, miles away, because he could not subsist it at the danger point. Late in March Pendleton found it impossible to bring up the horses for the artillery because they could not be fed, and reported that the artillery must be reduced because of the lack of horses.[235] When a few days later Sheridan's heavy force crashed through the weak Confederate right flank at Five Forks, there was only a small cavalry force left to oppose him. Lee later attributed the disaster in part to the absence of the cavalry units which had been sent to the interior to winter their horses and had not rejoined the army.[236]

With his flank turned and his remaining communications about to be cut, Lee began at once the withdrawal which he had long foreseen must be made. It would have been a difficult operation with his animals in

---

[235] Pendleton to W. H. Taylor, March 18, Pendleton to Chew, March 20, 1865, *Offic. Rec.,* ser. I. v. XLVI., pt. 3, pp. 1322, 1327.

[236] "The absence of the troops which I had sent to North and South Carolina, was I believe, the cause of our immediate disaster. Our small force of cavalry (a large portion of the men, who had been sent to the interior to winter their horses, had not rejoined their regiments,) was unable to resist the united Federal Cavalry, under Sheridan, which obliged me to detach Pickett's Division to Fitz Lee's support, thereby weakening my main line, and yet not accomplishing my purpose. If you had been there with all of our cavalry, the result at Five Forks would have been different. But how long the contest would have been prolonged, it is difficult to say. . . . " R. E. Lee to Wade Hampton, August 1, 1865, printed in Wade Hampton, *Address on the Life and Character of Gen. Robert E. Lee,* etc., p. 45 (Baltimore, 1871).

good condition; but now at the end of a severe winter when they were weak and slow from exposure and starvation it was a desperate undertaking. Only the stronger teams were able to take out wagon trains and guns, and on the forced marches without food they soon broke down. The cavalry could not keep pace with the better horses of Sheridan. At the end of a week what was left of a proud army was surrounded and the long struggle was over.

The University of Texas.
Charles W. Ramsdell

The Confederate Government and the Railroads[237]

by

Charles W. Ramsdell

For more than a year before the end came the railroads were in such a wretched condition that a complete breakdown seemed always imminent. As the tracks wore out on the main lines they were replenished by despoiling the branch lines; but while the expedient of feeding the weak roads to the more important afforded the latter some temporary sustenance, it seriously weakened the armies, since it steadily reduced the area from which supplies could be drawn.

The history of the Southern Confederacy affords an excellent illustration of the handicaps which, in this modern industrial world, beset any purely agricultural people in waging war. Success in war now depends so much upon the effective organization and application of

[237] Charles W. Ramsdell, "The Confederate Government and the Railroads", *The American Historical Review*, Vol. 22, No. 4 (Jul., 1917), pp. 794-810. Article and citation are verbatim. This paper was read at the meeting of the American Historical Association in Cincinnati, December 27, 1916.

the industrial resources of the nation to the support of the army that the mobilization of mines, farms, factories, foundries, banks, and means of transportation must accompany the mobilization of men. And, just as a trained army cannot be created without trained officers, the resources of a nation cannot be organized for effective military use if there is no body of trained industrial officers to conduct the industrial mobilization. When a people in a primitive stage of industrial development and therefore without trained industrial leaders engages a powerful adversary who is abundantly supplied with them, the tragic ending of that encounter is easily foretold. It was into such a conflict that the South rushed so light-heartedly in 1861.

Much has been said and written of the inferiority of the South in the supply of men and guns, when in fact a more fundamental weakness was its backward industrial condition. Moreover, industrial inexperience strengthened the confirmed particularism of the Southern people and their deep-rooted suspicion of every proposition which involved the extension of the activities and powers of the general government into the field reserved by custom for private enterprise. It would not be difficult to show that these were potent causes of the administrative paralysis which prostrated the Confederacy as much as did the battering of the Federal armies. Of this general statement the history of the Southern railroads from 1861 to 1865 offers one of the best illustrations.

The American Civil War was the first great military conflict in which railways were a highly important factor. So vast and in many parts so thinly populated was the area over which operations must be conducted and from which supplies must be drawn that without railways it would have been impossible for either side to maintain large armies at the front unless within reach of water transportation.[238] Even in the North where the railroads were better developed and the total mileage was twice that of the South, and where the Ohio, the upper Mississippi, the Potomac, and the sea furnished effective supplement to the roads—even there the problem of transporting men and supplies to the military frontier was a troublesome one. In the South, where the roads were in most cases short local lines, inadequately financed by local capital, cheaply constructed, poorly equipped, and supplemented but very little by water navigation, they were wholly unprepared for the task suddenly forced upon them by the war.

From the utter absence of any recorded discussion of the subject it is clear that at the outbreak of war no man of prominence in the Confederacy foresaw that the railroads were to play a part of great importance or that there was any urgent need of strengthening them. Upon the railroad companies themselves the first effects of the war were unfortunate. The business depression which came with hostilities, the establishment of the blockade, and the discouraging by the Confederate government of the exportation of cotton had greatly and suddenly reduced seaward traffic and revenues. Not knowing what was ahead of them the companies reduced expenses. Salaries were cut and trains and employees were laid

---

[238] *Cf.* Pratt, *Rise of Rail-Power in War and Conquest, 1833-1914,* p. 14 ff.

off.[239] Many of these employees were skilled men who were permanently lost to the roads, for some went into the army while others, of Northern birth and sympathies, made their way out of the Confederacy. Although the roads were now cut off from the Northern foundries from which they had always obtained their rails and rolling-stock, no general effort seems to have been made to get supplies elsewhere—a negligence which was probably due to the belief that the war would not last long.

Traffic soon revived but in a new direction. While the lines leading only to the sea-coast vainly awaited the raising of the blockade and the revival of business in the fall, those leading to the Virginia and Tennessee frontiers, where troops and stores were being concentrated, were enjoying a government patronage which greatly exceeded their former business. But the situation had its difficulties. The railroad business was still in the competitive stage in 1861 and the immense patronage of the government was worth fighting for; but as no single line could control rates all the way through from the lower South to Virginia and therefore none had much to gain by cutting its own rates, while it might be seriously affected by the combined rates of other roads, it is not surprising that the railroad men early came to the conclusion that some effort should be made to establish uniform charges for government transportation. This was desired also by the quartermaster-general, whose duty of providing all military transportation would be greatly lightened by such an arrangement. In the first flush of warlike

---

[239] Cuyler, Report of the President of the Southwestern Railroad, *Savannah Republican,* August 14, 1861. —AM HIST. REV., Vol. XXII.—51.

enthusiasm some of the roads had offered their services free for military purposes,[240] while others had charged their full local rate. Manifestly, this could not continue. Therefore the representatives of thirty-three roads met in convention at Montgomery on April 26, 1861, and agreed to a uniform rate of two cents a mile for men and half the regular local rate for munitions, provisions, and material, and also agreed to accept Confederate bonds at par in payment of government transportation.[241] Since the local rates varied greatly, this arrangement did not give complete satisfaction, and the railway presidents held another convention at Chattanooga on October 4, 1861, at which a schedule was drawn up and presented to Quartermaster-General A. C. Myers. This schedule divided the freights into four classes with a uniform rate of so much per one hundred miles for each class. After some consideration Myers accepted it and urged it upon the roads not represented at Chattanooga.[242] Although this rate schedule remained in force for some time, various roads on one pretext or another demanded a higher rate, which was in some cases granted, with the result that new uniform rates became necessary.[243] As the currency depreciated, higher and higher rates were

---

[240] *Official Records, War of the Rebellion,* fourth series, I. 120, 224, 236, 238.

[241] Quartermaster-General's Letter-Book, I. 98-1oo, Confederate Archives, U. S. War Department; *Offic. Rec.,* fourth series, I. 269, 538.

[242] Circular of Quartm.-Gen., December 13, 1861, Letter-Book, II. 442.

[243] A. C. Myers to various persons, Quartm.-Gen's. Letter-Book, IV. 232; VI. 6, 77, 117, 227 , 278, 301.

authorized in special cases to the end of the war.[244] The government never attempted to fix or interfere with rates for private business, probably assuming that such action was beyond its constitutional powers.

When in May and June of 1861 the government began to collect an army in Virginia one serious weakness of the transportation system was distinctly revealed. At such points as Chattanooga, Knoxville, Bristol, Lynchburg, Savannah, Augusta, Charlotte, Raleigh, Wilmington, and Petersburg—and there were many others—the roads terminating in those towns did not connect with each other and freight must be unloaded at one depot, hauled across town, and reloaded on cars at the other. Passengers frequently had to wait over until the next day. Since this arrangement made business for hotels and transfer companies, the town looked upon it with favor as a valuable asset and strongly opposed every attempt to provide connections for through traffic.[245] Even where the tracks connected, the freight had to be unloaded and reloaded on other cars, since no company was willing to entrust its cars to another line. Frequently troops and stores so unloaded would be compelled to wait for days and even weeks before they could move on to the next terminus. Consequently, at these points of congestion troops, ordnance, quartermasters, and commissary stores began to accumulate, and confusion, further delay, and sometimes heavy losses resulted.[246]

---

[244] W. S. Alexander to James Stewart, January 14, 1864, Quartm.-Gen's. Letter-Book, VII. 528; H. K. Burgwyn to J. A. Seddon, *Offic. Rec.,* fourth series, III. 616-618.

[245] *Savannah Republican,* November 11, 1861, for conditions at Augusta.

[246] Myers to J. S. Barbour, June 17, 1861, Quartm.-Gen's.

Steps were taken to bridge these gaps, but without much effect. The case of Petersburg, Virginia, may be taken as an example. So great was the delay, expense, and inconvenience of transshipment between the several roads terminating at that important point that Gen. Robert E. Lee at the very beginning of the war urged the construction of connecting tracks.[247] The railway companies had long desired to make the connection but had been prevented from doing so by the opposition of the town itself. As a question of law was involved the Virginia state convention passed an ordinance, June 26, 1861, authorizing the connection. The roads now asserted that the expense would be too great for them to undertake the work without government loan; and when this seemed in a fair way to be obtained, the question arose whether the new law contemplated a permanent connection, in which the railroads would have an interest, or only a temporary one in which they would have practically none. As a temporary connection would be of light and flimsy construction and impassable for heavy freights, and as the authorities of Petersburg continued strenuously to oppose a permanent connection, no action was taken, and the congestion continued.[248] At other connecting points,

---

Letter-Book, I. 197-203; Myers to Campbell Wallace, July 9, 1861, *ibid.,* p. 275; Myers to G. R. Echols, July 12, 1861, *ibid.,* p. 287; Myers to M. J. Hannan, July 12, 1861, *ibid.,* 287; Myers to E. H. Gill, July 22, 1861, *ibid.,* p. 322.

[247] Lee to E. T. Morris, June 18, 1861, *Offic. Rec.,* fourth series, I. 394.

[248] P. V. Daniel, jr., to .Davis, June 27, July 17, 1861, *Offic. Rec.,* fourth series, 405, 484; Daniel to Walker, July 2, *ibid.,* p. 417; Wm. T. Joynes to Walker, July 17, *ibid.,* p. 485; "Resolutions of the Common Council of Petersburg", December 10, 1861, *Public Documents of the Virginia*

as Lynchburg, through traffic was impossible because of the change of gauge. The confusion and congestion were not relieved by the frequent interference of lesser military officials, who sought on their own responsibility and regardless of schedules and distribution of rolling-stock to order trains back and forth within the limits of their respective commands.[249] The quartermaster-general had no control over these officers and he was put to the utmost exertions to straighten out the tangles and mollify the railroad officials. The first general order issued by General Lee after he was called to Richmond in March, 1862, was directed against this practice.[250] It was becoming increasingly evident that some system of effective supervision or control looking to better co-ordination of shipments would soon be essential to the supply of the armies and the safety of the government itself.

Another difficulty which appeared early and steadily grew worse under the stress of war was the shortage of cars and engines. The supply on most of the Southern roads was scanty before the war and it proved wholly inadequate for the needs of the government. Moreover, some of the roads upon which the heaviest traffic was thrown were least able to bear it. This was the case with the line of roads extending from the vicinity of Chattanooga up the Tennessee valley and across to Lynchburg. These roads were comparatively new and

---

Assembly, 1861-1862, no. 32.

[249] Wallace and John R. Branner to Benjamin, December 4, 1861, *Offic. Rec.,* first series, vol. LII, pt. 2, pp. 227-228; Myers to Joseph E. Johnston, March 6, 1862, Quartm.-Gen's. Letter-Book, III. 380.

[250] See *Offic. Rec.,* fourth series, I. 1010-1011.

their traffic before the war had been light, but now they became the chief carriers of grain, beef , and pork from the Tennessee region to the armies in Virginia. The task was far beyond their capacity and the continuous use of cars and engines without giving time for repairs reduced both rolling-stock and the frail tracks to a sad condition. The quartermaster-general made repeated efforts to obtain cars and engines from other roads for use on this line; but, as this aroused the jealousy of the officials of the other roads, who protested vigorously that they had none to spare, little was procured even under threat of impressment.[251] In fact every road was suffering for cars and engines, since the few shops which would build or repair them were soon either leased by the government for its own uses or were crippled by the conscription of their skilled workmen. It was becoming impossible to replace losses and by the opening of 1862 the shortage of rolling-stock was so alarming that railroad men were predicting the utter breakdown of the roads within a short time.[252]

---

[251] Myers to Wallace, September 18, 1861, Quartm.-Gen's. Letter-Book, II. 60; Benjamin to Myers, September 24, and Myers to W. S. Ashe, September 25, 1861, *Offic. Rec.,* fourth series, I. 617; Myers to Wallace, September 30, Quartm.-Gen's. Letter-Book, II. 103; Myers to Ashe, October 5, 1861, *ibid.,* p. 123; Benjamin to Joseph E. Brown, September 30, and Brown to Benjamin, October 2 and 4, 1861, *Offic. Rec.,* fourth series, I. 634, 646, 666.

[252] Neill S. Brown to Benjamin, January 12, 1862, *Offic. Rec.,* fourth series, 839; resolutions of a convention of railroad presidents, Richmond, December 6, 1861, Pickett Papers, Library of Congress, accession 1910, fol. 108; Myers to H. J. Ranney, January 10, 1862, Quartm.-Gen's. Letter-Book, III. 106-107.

Imbued as the Southern people were with *laissez faire* ideas, their government was slow to take a hand in the operation of the roads, and when finally compelled by force of circumstances to interfere, it came only by degrees to any assumption of control. When the congestion of traffic in the summer of 1861 became serious, W. S. Ashe, formerly president of the Wilmington and Weldon, was appointed major and assistant quartermaster and assigned to the duty of "superintending the transportation of Troops and Military stores on all the Railroads, North and South, in the Confederate States". He was directed to give his special attention to the detention of freights on the roads from Wilmington to Richmond and from Nashville to Richmond, and to obtain concert of action among the several roads in order "to control the movement, speed, time-table, and connections" of the numerous trains going out of Richmond.[253] How long Ashe was retained in this position is not clear, nor is the exact extent of his authority anywhere defined. His rank and the correspondence of the quartermaster-general with various railroad officials indicate that Myers kept the general control of the business in his own hands and that he employed Ashe only as a sort of traveling agent and inspector to make contracts, investigate complaints, give assistance to the roads where possible, and make recommendations to the quartermaster-general.[254]

This arrangement seems to have accomplished but

[253] Myers to Ashe, July 18, 1861, Quartm.-Gen's. Letter-Book, I. 313.

[254] Quartm.-Gen's. Letter-Book, I. 322, 332; II. 103, 123, 187, 353, 442; Ashe to Davis, November 27, and December 13, 1861, Pickett Papers, accession 1910, fol. 108.

little toward the solution of the problems and was evidently unsatisfactory to the Secretary of War, by whose order Col. William M. Wadley, president of the Vicksburg and Shreveport Railroad, was, on December 3, 1862, assigned to the "supervision and control of the transportation for the Government on all the railroads in the Confederate States".[255] Wadley's powers were somewhat more extensive than those previously assigned to Ashe, especially in respect to control over government agents, employees, engines, cars, and machinery. He was further not to be subject to the orders of the quartermaster-general but was to report through the adjutant and inspector-general to the Secretary of War. Of this last provision Myers complained repeatedly that only inconvenience, confusion, and embarrassment could result from transferring to another division of the war office the supervision of a service for which the quartermaster-general was responsible.[256] Myers was not in the good graces of the administration now, however, and his protests were unheeded.

Colonel Wadley tried to induce the railroad heads to agree (1) to a definite plan of co-operation under his immediate supervision, by which each railroad superintendent should act as an assistant to him and make weekly reports, and (2) to a through schedule of

---

[255] "General Orders, No. 98, Adjt. and Insp. General's Office", *Offic. Rec.,* fourth series, II. 225. Wadley had been in the railroad business for many years in Georgia and Mississippi. Phillips, *History of Transportation in the Eastern Cotton Belt,* p. 319; Pollard to Davis, April 4, 1862, *Offic. Rec.,* fourth series, I. 1048.

[256] Myers to Seddon, December 9, 1862, January 8 and 26, 1863, *Offic. Rec.,* fourth series, II. 231, 304, *372.*

trains between Montgomery and Richmond. But his efforts were without success. The roads on the contrary adopted a schedule of rates which Wadley considered inequitable.[257] It seems that he never acquired direct control over any of the roads further than allowed by the contracts he was able to induce them to enter into, and his activities were confined chiefly to settling rates, helping the destitute roads to obtain rolling-stock, and making recommendations to the Secretary of War for the assumption by law of direct control and management of the roads that failed to perform their full duty.[258] For some reason not disclosed, Wadley's nomination was not agreeable to the Confederate senate and was rejected May 1, 1863.[259] Thereupon Capt. F. W. Sims was appointed June 4, 1863, to his position with the same duties and powers.[260]

On August 10, 1863, Brig.-Gen. A. R. Lawton replaced Colonel Myers as quartermaster-general, and the railroad bureau, of which Sims was the head, was placed at once under his jurisdiction.[261] Sims was both an able and an industrious officer and strove hard to improve the condition of the railroads, but his efforts to have detailed from the army enough mechanics to set the shops to building cars and engines and to rolling rails had no effect upon the higher military authorities. He seems to have

---

[257] Wadley to Cooper, December 31, 1862, *ibid.*, pp. 270-278.
[258] Wadley to Seddon, January 26, April 14 and 15, 1863, *ibid.*, pp. 373, 483, 486; Myers to Larkin Smith, April 23, 1863, Quartm.-Gen's. Letter-Book, VI. 301.
[259] *Journal of Congress of Confederate States*, III. 426.
[260] *Offic. Rec.*, fourth series, II. 579.
[261] *Ibid.*, p. 697; Lawton to Sims, August 12, 1863, Quartm.-Gen's. Letter-Book, VII. 3 1.

had the confidence of the railroad men, probably because he showed a sympathetic understanding of their difficulties. Sims retained his post until the end of the war, but during the last year the duty of repairing the roads, especially bridges, was imposed upon the engineer bureau; and as the duties of supervision became too heavy for one man, the quartermaster-general from time to time called upon experienced railroad men in distant parts of the country to take charge of the transportation in those regions. [262]

Mere supervision could not make the transportation system efficient. Early in the war it became clear that the roads could not unaided procure the supplies and repairs necessary to keep them in good condition, and it was not long before they turned to the government as the only possible source of help. Besides, new lines were needed to link together certain neighboring roads in order to shorten distance and both to cheapen and to expedite shipments; and the building of these connections would require a financial backing which only the government was able to give.

The most important connection proposed was that between Danville, Virginia, and Greensborough, North Carolina. Mr. Davis called attention to the advantages of bridging this gap of about forty-eight miles, in his message to the Provisional Congress in November, 1861.[263] It was estimated that a loan of one million dollars would be sufficient to provide for the speedy construction of the road, and Congress passed an act on February 10,

---

[262] Lawton to Sam Tate, October 6, 1863, *ibid.*, p. 178; Lawton to Thomas Peters, November 27, 1863, *ibid.*, p. 364.

[263] *Journal of Cong. of Conf. St.*, I. 470.

1862, authorizing the loan.[264] Mr. Davis had expressed the opinion that since the work was "indispensable for the most successful prosecution of the war, the action of the Government will not be restrained by the constitutional objection which would attach to a work for commercial purposes". Some of the foremost members of Congress thought differently and fought the bill with every available resource; and after its passage they caused to be spread upon the journal a protest against the act as an unwarranted and dangerous violation of the constitution under the guise of military necessity.[265] The actual construction of the road was delayed for more than two years, partly by the necessity of completing satisfactory surveys and examining rival routes, partly by the scarcity of labor and material.[266] Connection was established about May 20, 1864.[267] Though flimsy of construction and prolific of wrecks, this road, opened just after the beginning of Lee's desperate struggle with Grant, was of great benefit to the Confederates and became more and more important when later in that year the Weldon railroad was threatened. Another important connection, which was undertaken at about the same time, was that between Meridian, Mississippi, and Selma, Alabama. This

---

[264] *Statutes at Large of the Provisional Congress,* p. 258.

[265] *Journal of Cong. of Conf. St.,* I. 731-734, 762-764, 766-770, 781-782. Among the ten who signed the protest were Robert Toombs, who was evidently its author, R. B. Rhett, J. L. M. Curry, W. S. Oldham, and M. J. Crawford. A. H. Stephens voted against the bill but did not sign the protest.

[266] *Offic. Rec.,* fourth series, I. 947, 1025-1027, 1085-1087; III. 392-393.

[267] Lawton to Chisman, May 19, 1864, Quartm.-Gen's. Letter-Book, VIII. 237.

would not only greatly shorten the route from Richmond to Vicksburg and New Orleans, but by giving Vicksburg direct communication with central Alabama and Georgia would greatly strengthen that important post. The distance was about one hundred miles, but for about half of this distance a road had already been completed and part of the rest was graded. The company sought an advance of $150,000 from the government and Mr. Davis recommended it to Congress.[268] An act of February 15, 1862, carried out the recommendation, but it soon was discovered that the sum was insufficient because of the rapidly increasing cost of materials. After many delays this road was completed about the end of 1862.[269] While the last-mentioned bill was before Congress a third was introduced, to lend money to establish a connection between the roads in western Florida and southwestern Georgia, but it failed on third reading.[270] In April, 1862, the first permanent Congress authorized a loan of $1,500,000 to aid the construction of a line between New Iberia, Louisiana, and Orange, Texas, which would give direct railway connection between Houston and New Orleans and make the resources of Texas available for the defense of the lower Mississippi. The fall of New Orleans shortly afterwards rendered the prosecution of the work

---

[268] Ashe to Davis, November 27 and December 13, 1861, Pickett Papers, accession 1910, fol. 108; *Journal of Cong. of Conf. St.,* I. 586.

[269] James L. Price to Randolph, April 10 and 15, 1862, *Offic. Rec.,* fourth series, I. 1053, 1060; Gaines to Randolph, April 24 and June 25, 1862, *ibid.,* pp. 1089, 1171; Shorter to Randolph, October 27, 1862, *ibid.,* II. 148-149.

[270] *Journal of Cong. of Conf. St.,* I. 819, For the genesis of the bill, see *Offic. Rec.,* fourth series, I. 612, 777-779.

useless and it was abandoned.[271] In October, 1862, Congress passed an act which authorized the President to cause a railroad to be constructed between Rome, Georgia, and Blue Mountain, Alabama, and appropriated $1,122,480.92 for that purpose.[272] Such a road would not only establish a new connection from northern Georgia through central Alabama to the Mississippi, but, more important, would give access to the great iron and coal deposits in Alabama. After much delay, because of the difficulty in procuring iron, construction was begun, but the road was not completed before the end of the war.[273] All of these acts were based upon "military necessity" and all of them were steadily opposed by the ultra-conservative strict-constructionist minority in Congress. Numerous other railway companies made appeals for aid, but no action was taken in their behalf until the beginning of 1865 when, upon the recommendation of the Secretary of War, Breckinridge, and the President, a blanket appropriation was made, March 9, for the construction and repair of railroads for military purposes.[274]

Where and how to procure material for laying tracks, building bridges, and constructing or repairing engines

---

[271] *Ibid.*, p. 1013 *Journal of Cong. of Conf. St.*, I. 238, 361; V. 260, 261, 279; II. 195, 197-199; *Offic. Rec.*, fourth series, I. 1073; II. 107-108.

[272] *Ibid.*, II. 200-201.

[273] Shorter to Davis, October 25, 1862, *ibid.*, pp. 144-146; Campbell to Bragg, April 21, l864, *ibid.*, III. 312; Fleming, *Civil War and Reconstruction in Alabama*, p. 156.

[274] *Offic. Rec.*, fourth series, III. 1095-1096, 1114; *Journal of Cong. of Conf. St.*, VII. 671, 685, 709, 749. The amount appropriated is not shown in the journal, but $21,000,000 was recommended. This, however, was in greatly depreciated currency or bonds.

and cars was the most difficult problem of the railroads and it was fundamental to their very existence. Iron and machinery were especially scarce. Before the war these necessities had been supplied from the North: now they must be manufactured or imported from Europe. But few iron mines, smelters, and foundries existed in the South in 1861, and these were small and were soon under contract to their full capacity with the ordnance department. It seemed that if railway foundries or machine shops were necessary, the roads themselves must build them. But it was a serious question whether the average company could afford to build a complete set of shops of its own when it operated only a short line of one hundred to two hundred miles—and some of the most important were even much shorter. The capital required was out of proportion to the size and earning capacity of the road. Moreover, because of the widespread belief that the war would be short, there was at first a natural reluctance to invest large sums in plants which would almost certainly prove unprofitable after the coming of peace.[275] For a group of roads to combine for the purpose involved practical difficulties of management which they were unprepared to solve or even to attempt. Besides, it was clear that however well supplied with machine shops, the roads would still be helpless so long as the government monopolized the output of iron and continued to conscribe skilled workmen into the army.[276]

For these reasons and because the building of shops

---

[275] Sims to Lawton, October 23, 1863, *Offic. Rec.,* fourth series, II. 881.

[276] *Offic. Rec.,* fourth series, II. 881; also, Sims to Lawton, February 10, 1865, *ibid.,* III. 1092.

and mills, even if determined upon, would take time, the railroad men at first tried to import supplies from Europe. But the growing stringency of the blockade and the lack of well-established commercial or credit relations with European firms made this very difficult. The administration refused to take any part in promoting or financing large mercantile combinations for the purpose of establishing credit accounts in Europe based upon cotton,[277] and the roads were unable to command enough capital, or cotton, and steamboat transportation, independent of government aid, to make importations on their own account. Nor would the government, though frequently appealed to, itself import railroad material for sale to the roads.[278] On one occasion certain Virginia roads were allowed to purchase supplies in England through an agent of the War Department; but the favor was not allowed again, the secretary, Mr. Seddon, explaining that he was unwilling to intervene officially in matters relating "exclusively" to the interest of the railroads.[279] In 1864 the president of a railroad in Mississippi, which was cut off from the sea, obtained military permission to export cotton through the Federal lines and to bring in railroad material in return, but this

---

[277] For one proposition of this character, see D. T. Bisbie to Benjamin, January 16, 1862, *ibid.,* I. 843-845; E. Fontaine to Benjamin, *ibid.,* p. 868.

[278] Resolutions of a railroad convention at Richmond, December 6, 1861, Pickett Papers, accession 1910, fol. 108; Daniel to Seddon, February 12, 1863, *Offic. Rec.,* fourth series, II. 394; also April *22,* 1863, *ibid.,* pp. 499-510.

[279] Seddon to J. M. Robinson, February 24, 1863, *ibid.,* p. 409; Daniel to Seddon, April 23, 1863, *ibid.,* p. 511; September 30, *ibid.,* pp. 841-842; Seddon to Daniel, October 3, 1863, *ibid.,* p. 852; Daniel to Seddon, October 9, *ibid.,* p. 866.

was exceptional.[280] The total amount of railroad supplies brought in from outside the Confederacy was trifling. If the needed materials were to be had there was only one way left: they must be produced in the Confederacy and the government must aid directly in the work. Early in the war some of the more far-sighted railroad men had pointed out the possibilities of utilizing and improving the railroad shops with government aid and the advantages to the government of contracting for cars and engines of its own.[281] The administration, however, for the time preferred to contract with the companies directly for transportation and to leave to them the problem of maintaining the efficiency of their roads. Before 1863 the entire output of most of the shops, foundries, mines, and mills was absorbed by government contracts and except in a few cases—chiefly in Georgia—the roads were without any means of manufacturing even the simplest materials. When the rolling-stock on one road wore out, the transportation officers sought it from other roads and in many cases impressed it. When this hand-to-mouth policy failed, the quartermaster-general contracted for the building of cars for government use; but he was never able to obtain enough, for not only was material lacking but sufficient details of mechanics could not be obtained from the army to carry out any large contract. Special agents and commissioners were detailed to inspect roads, impress, collect, and redistribute rails; and the smaller

---

[280] *Ibid.,* pp. 381, 388; III. 478, 508, 514, 65 I.

[281] Goodman to Davis, January 25, 1862, *ibid.,* I. 880-882; Neill S. Brown to Benjamin, January 12, 1862, *ibid.,* p. 839; Daniel to Seddon, April 22, 1863, *ibid.,* II. 499-508, 511; Cuyler to Seddon, April 22, 1862, *ibid.,* pp. 508-510; Jones, *A Rebel War Clerk's Diary,* I. *302* (April 30, 1863).

and less important roads were stripped of both rails and rolling-stock to keep the main lines in operation.[282] Lieutenant-Colonel Sims, the superintendent of railroad transportation, made repeated appeals for government aid in the manufacture of supplies, for "men and iron", but without substantial effect; and on February 10, 1865, we find him lamenting that "not a single bar of railroad iron has been rolled in the Confederacy since [the beginning of the war, nor can we hope to do any better during the continuance".[283]

Although forced by military necessity to interfere frequently with the operation of the roads and to exercise an ever-increasing control over them, the Confederate government disclaimed any intention of doing more than to compel the railroad officials, under the contracts which had been made, to give priority to government freight over that of private persons and to expedite shipments.[284] This exaction became a matter of serious concern both to the roads and to private shippers as the traffic grew heavier and the roads weaker. By the end of 1863 there was no room except at intervals for anything but government freight on the main lines. Since Virginia and

---

[282] Seddon to Kenney, July 21, 1863, *Offic. Rec.,* fourth series, II. 655; "Special Orders No. 232", September 30, 1864, *ibid.,* III. 694; J. F. Gilmer to Breckinridge, February 16, 1865, *ibid.,* p. 1085; Sims to Lawton, February 10, 1865, *ibid.,* pp. 1091-1093; Wallace to Gov. Z. B. Vance, February 6, 1863, Vance's Letter-Book, I. 124, in Confederate Archives, U. S. War Department; Cowan to Vance, September 5, 1864, *ibid.,* pp. 568-570.

[283] *Offic. Rec.,* fourth series, III. 93, 229, 1092.

[284] Randolph to Shorter, November 8, 1862, *ibid.,* II. 175; Myers to Joseph E. Brown, February 11, 1863, Quartm.-Gen's. Letter-Book, VI. 6.

North Carolina had been stripped bare of provisions, Lee's army was now being supplied from South Carolina and Georgia and the roads to the south of Richmond were overworked. In March, 1864, all passenger trains in North Carolina were stopped for several days to permit the passage of corn to Richmond.[285] The order raised a storm of protests, but the quartermaster-general and the Secretary of War held to it until the stores of corn were brought up. The stopping of passenger trains became a frequent occurrence thereafter and private travel along the roads to Virginia practically ceased. Some communities along the Wilmington and Weldon road were threatened with actual famine because the War Department would not relax the rule of priority in order that they might bring in their own supplies of corn.[286] The rule probably was not enforced with absolute rigidity, because we find that station quartermasters are frequently charged with violating it, and in fact it was claimed that speculators, by the aid of bribes, could usually get their shipments through. All trains running to Wilmington—which was the only seaport left in 1864—were required to take government freight, usually cotton, to at least half of their capacity.[287] The control of transportation gave the quartermaster's department a powerful weapon with which to force manufacturers of cotton goods, especially in North Carolina, to make

[285] Lawton to Vance, Cameron, Echols, and Lee, March 11, 1864, Quartm.- Gen's. Letter-Book, VIII. 91; and to various others, March 16, 18, 30, and April 7, 12, *ibid.,* pp. 97, 101, 131, 152, 160.
[286] Lawton to W. A. Graham, June 8, 1864, *ibid.,* p. 277.
[287] Lawton to J. W. Cameron, December 11, 1863, *ibid.,* VII. 420; Lawton to Davis, September 20, 1864, *ibid.,* IX. 129.

contracts to furnish the government with cloth—usually at prices below the market rate—for without the consent of the department they could obtain no shipments of raw cotton.[288]

Time-honored conventional theories about the limitation of the functions of government had begun even early in the war to give way before the pressure of imperious military necessity. As the responsible government officials had been led to interfere more and more in railroad affairs in order to sustain the armies at the front, it became increasingly evident that the railroad companies, if left to themselves, either could not or would not render the service which the government must have. while most of the obstacles to efficiency lay in conditions for which the roads were not to blame, there were others which the railroad officials themselves seemed to raise. But in some cases it is not easy to censure them. For instance, when every company was trying anxiously to husband its scanty supply of rolling-stock it was only natural that it should resist every attempt to have its cars run on through to Richmond or other distant destination and should insist on breaking bulk and reloading at its own terminus.[289] And it is hardly surprising that the commission of army officers appointed to collect and redistribute railroad iron should find "every possible impediment" thrown in its way and its efforts often

---

[288] Lawton to W. G. Ferguson, September 12, 1864, *ibid.*, p. 96.

[289] In respect to this the roads were sustained by both Colonel Sims and General Lawton. See *Offic. Rec.*, fourth series, III. .228; and Lawton to Davis, September 20, 1864, Quartm.-Gen's. Letter-Book, IX. 129.

defeated.[290] As the freight rates paid by the government were far below those for private business, railroad officials connived with civilians to forward freight for the latter even at the cost of holding up army supplies. Disagreements between the various companies frequently caused needless delays and unnecessary diversion to roundabout routes. Those roads which were owned in whole or in part by states were especially troublesome because they took refuge behind the authority of the state. The most conspicuous was the Western and Atlantic of Georgia, championed by Gov. Joe Brown; but certain roads in North Carolina and Florida also took a very independent course.[291] In all these cases the Confederate authorities found themselves helpless because the government was unwilling to incur the odium of overriding state authority.

The first quartermaster-general, Myers, had steadily opposed the assumption of governmental control over the railways.[292] It seems that when Sims became head of the railroad bureau in 1863 he held similar views, but by

---

[290] Report of Gilmer to Breckinridge, February 16, 1865, *Offic. Rec.,* fourth series, III. 1085.

[291] For correspondence relative to condition of the Western and Atlantic Railroad see *Offic. Rec.,* first series, vol. LII, pt. 2, pp, 593, 596, 601, 607-608, 616, 621-623; vol. XXXII., pt. 2, pp. 591; for Florida Railroad, *ibid.,* LIII. 350-359, 362-364; for North Carolina roads, Cowan to Vance, February 13, 1864, Vance's Letter-Book, I. 458-462; Vance to Gilmer, *ibid.,* p. 552; Gilmer to Vance, *ibid.,* p. 561; Cowan to Vance, *ibid.,* pp. 568, 571; Cowan to Gilmer, *ibid.,* p. 570; Vance to Gilmer, *ibid.,* p. 572.

[292] Myers to Davis, January 31, 1862, *Offic. Rec.,* fourth series, I. 896; Myers to Chilton, October 3, 1862, *ibid.,* II. 108.

April, 1864, he said:

> That the railroads should come under military control I am becoming every day more satisfied. There seems to be a desire to work for the road's interest rather than sacrifice all convenience for the country's cause. . . . Greater harmony would doubtless produce better results, but this I fear can never be obtained until a Government officer manages every road.[293]

President Davis probably never seriously considered the idea of subjecting any of the railroads, except in a military exigency, to complete control by the government. His few references to the subject in his messages to Congress are almost casual and indicate that beyond delegating a general supervision of the government's interests to Wadley, Sims, and the quartermaster-general, to be enforced by threat of impressment, he was unprepared to go.[294] Nor was the majority of Congress at first willing to go very far in this direction. On .August 21, 1861, a bill was introduced from the committee on military affairs "authorizing the President to regulate and take control of railroads in certain cases", but it failed to become a law.[295] In January, 1862, a special committee which had been appointed to examine into the various divisions of the

---

[293] Sims to Lawton, April I, 1864, *Offic. Rec.,* fourth series, III. 228.

[294] Richardson, *Messages and Papers of the Confederacy,* I. 139, 152, 236, *295.* See also report of Secretary of War (Seddon), January 3, 1863, *Offic. Rec.,* fourth series, II. 293.

[295] *Journal of Cong. of the Conf. St.,* I. *379.*

War Department recommended, among other things, that military control be taken of all railroads terminating in or passing through Richmond, Nashville, Memphis, and Atlanta, or leading to the headquarters of the several army corps, and that they be placed under an efficient superintendent.[296] This recommendation was without immediate result, but in the first session of the permanent Congress the House of Representatives passed a bill "to provide for the safe and expeditious transportation of troops and munitions of war by railroads"[297] It directed the President to appoint "a military chief of railroad transportation", who should be selected from the railroad officials, and provided that the regular officials of each road should be given a stated rank, should be left as far as possible in charge of their own roads, but be subject to the orders of the chief and liable to court martial for neglect of duty.[298] The bill was referred in the Senate to the Committee on Military Affairs, was reported without amendment on April 21, the last day of the session, and failed for want of time to come up for passage. Although other bills of this general character were introduced, none became law until February, 1865, when finally an act was passed which authorized the Secretary of War to place any railroad, canal, or telegraph line under such officers as he should designate, to place the regular railroad officials, agents, and employees under these officers on the same footing

[296] *Ibid.*, p. 721.

[297] *Ibid.*, V. 251-254. For earlier suggestions see *ibid.*, p. 122, and II. 87 (Senate).

[298] For a protest against the bill, see *Journal*, V. 269. The Augusta *Chronicle and Sentinel*, April 24, 1862, endorsed the protest and denounced the bill as a "usurpation".

as soldiers in the field, and to maintain any road in repair or to give it any necessary aid. Provision was made for reimbursing the road for any damage sustained while in the hands of the government.[299] Whether this measure would ever have proved effective may be doubted, for it did not insure an improvement in the material condition of the roads, and death-bed resolutions are somewhat unconvincing.

For more than a year before the end came the railroads were in such a wretched condition that a complete breakdown seemed always imminent. As the tracks wore out on the main lines they were replenished by despoiling the branch lines; but while the expedient of feeding the weak roads to the more important afforded the latter some temporary sustenance, it seriously weakened the armies, since it steadily reduced the area from which supplies could be drawn. The rolling-stock, replenished in the same way, wore out so fast that some roads were nearly destitute of cars and engines. It is very difficult to get precise information about particular roads in 1864, but from scattered statements of Quartermaster-General Lawton it appears that on the roads from Georgia to Richmond not more than two or three trains per day could be run at the rate of one hundred miles per day and less. Despite the low rate of speed, accidents were frequent and helped to block the tracks. On one occasion Lawton declared it impossible for cars to be run through from Columbus, Georgia, to Charlotte, North Carolina, a distance of approximately five hundred miles, because

---

[299] *Journal,* VII. 584-587, 607, 707; IV. 571, 573-574. The act was approved February 28, 1865.

they would break down on the way if not repaired.[300] From the winter of 1863-1864 Lee's army had to draw its supplies from South Carolina and southern Georgia, a distance of from five hundred to nearly a thousand miles, and it rarely had more than two or three days' supply of food ahead. No surplus could be accumulated and as time wore on the supply became scantier. By the end of summer the roads could not bring enough, with the utmost exertions and even when unhampered by the enemy, to feed the men and horses half rations. Indeed it is hard to see how Lee could have maintained his army in Virginia for another year, even if Grant had been content to watch him peaceably from a distance. And yet Lee's army was starving not because there was no food in the Confederacy, for it was plentiful in many portions of Georgia, Alabama, and Florida, but because the railroads simply could not carry enough of it. Over and over again Lawton declares that "transportation is scarcer than provisions".[301] Corn brought in Richmond twenty and twenty-five times as much as it sold for in southwest Georgia.[302] When this region was cut off and the remnant of the feeble roads wrecked by Sherman's destructive march through Georgia and the Carolinas, the stoppage of all supplies followed, and the long struggle was over. It would be claiming too much to say that the failure to solve its railroad problem was the cause of the Confederacy's downfall, yet it is impossible not to conclude that the

---

[300] Lawton to Davis, September 20, 1864, Quartm.-Gen's. Letter-Book, IX. 129.

[301] Lawton to Maj. J. G. Michaeloffsky, Macon, Ga., January 19, 1864, Quartm.-Gen's. Letter-Book, VII. 543; Captain Seals, Fort Gaines, to McMahon, February 9, 1864, *ibid.,* VIII. 18.

[302] *Daily Examiner,* February 25, 1864.

solution of that problem was one of the important conditions of success.[303] The failure to solve it was due partly to the industrial unpreparedness of the South, partly to the shortsighted policy of leaving the task of maintenance entirely to the stockholders, although depriving them of the use of workmen and materials, partly to the apparent inability to comprehend the essentially public character and responsibility of the roads, and partly to an ingrained abhorrence of extending the activities of the general government into the field reserved to the states or to private enterprise. Had the Confederate government been able at the outset to adopt measures with respect to its railroads as vigorous and far-sighted as it did for its ordnance department, it seems certain that the roads would have been maintained and improved, and the effective resources of the Southern people and the strength of their armies would have been tremendously increased.

CHARLES W. RAMSDELL

---

[303] So keen an observer as William M. Burwell believed that the deficiencies in transportation aggravated the currency troubles. *Offic. Rec.*, fourth series, III. 226-227.

The Control of Manufacturing by the Confederate Government[304]

by

Charles W. Ramsdell

The military control which was gradually established was rigid, but partial, uneven, short-sighted. The sole purpose was to get supplies for the army at as reasonable a cost as possible—to exploit the factories, not to develop them for the benefit of the community at large. After all, these things were inevitable under a government struggling for bare existence and dependent upon undeveloped and inadequate resources.

The only manufactures over which the Confederate government sought control were those which directly supplied the needs of the army. These were of two classes: (1) arms and munitions, which were under charge of the

---

[304] Charles W. Ramsdell, "The Control of Manufacturing by the Confederate Government", *The Mississippi Valley Historical Review*, Vol. 8, No. 3 (Dec., 1921), pp. 231-249. Article and citation are verbatim. Punctuation and capitalization are exactly as is in original paper. This paper was read at a joint session of the American Historical Association and the Mississippi Valley Historical Association held in Washington, December 29, 1920.

ordnance bureau; and (2) a more diverse group which included clothing, blankets, tents, shoes, wagons, saddles and harness, which for the most part were provided by the quartermaster's bureau. Because of the limitations of space it has been necessary to restrict this paper to the consideration of the manufacture of certain items of the second group, namely clothing and shoes.

It is well known that manufactures of any kind had developed only to a very limited extent in the south before 1861. Textile factories had made a beginning, but they contributed but a small part of the cloth annually consumed. When the war came, the people at home revived the ancient household arts of spinning and weaving while the government was left to uncertain resources for the material with which to clothe and shelter the men it called to service in its armies. As no civil machinery existed for handling this problem, and the task of providing these supplies for the army devolved almost entirely upon the quartermaster 's bureau of the war department, it is with the activities of that organization that we shall be chiefly concerned.

Under the first act of congress that provided for a volunteer army, each soldier was required to furnish his own clothing, for which he was to be paid.[305] This was soon followed by a system of commutation by which the private soldier or noncommissioned officer was allowed a stated sum for the purchase of clothing.[306] Somewhat

---

305 "An act to provide for the public defense," approved March 6, 1861, Statutes at large of the provisional government of the Confederate states of America, 1 session, chapter 26, sections 3, 4.

306 "An act concerning transportation, etc.," approved May 21, 1861, *ibid.*, 2 session, chapter 39. The sum allowed was at

later, the secretary of war was required to provide clothing whenever the soldier was unable or unwilling to supply himself by means of the commutation.[307] This alternative system lasted for a little more than a year, until commutation was abolished by the act of October 8, 1862.[308] From this time on the war department was required to furnish all clothing to privates and noncommissioned officers. It had now become almost impossible for the soldier to provide his own clothing.

With the first establishment of army camps, quartermaster's supply depots were set up at convenient points for the reception, storage, and distribution of shoes, clothing, blankets, tents, and other necessities. Some of these articles were bought by the government, while others were contributed by the families of the soldiers or by patriotic organizations. Ready-made clothing was not then a market staple; therefore, when the quartermaster purchased cloth he must turn it over to some band of patriotic women who would volunteer to make it up into clothing or else he must hire sewing-women to do the work. The first method proved unsatisfactory, and by the fall of 1861 army clothing was being made almost exclusively in the quartermaster's depots under military supervision. This was plainly due to the absence of adequate privately owned facilities. As the depots were located in the larger towns where the cost of living was rising rapidly, it was easy to get plenty of labor, and thousands of women and girls sought employment in

---

first twenty-one dollars every six months. Later it was made twenty-five dollars.

[307] Act of August 30, 1861, *ibid.*, 3 session, chapter 51.

[308] *Ibid.*, 1 congress, 2 session, chapter 30.

the clothing shops. In some depots, as at Richmond and at Columbus, Georgia, shoe shops were set up beside the garment shops. In these the operatives were men.

These government shoe and clothing shops, usually referred to officially as "factories," continued operations until the end of the war, except where destroyed or withdrawn before the union armies, and in them were made practically all army shoes and clothing, except such as came through the blockade or were captured or were furnished by certain of the states to their own troops. Several of the shops were conspicuously well managed, especially those at Richmond, Augusta, Atlanta, and Columbus. Others were less successful, sometimes because of the inexperience of the officials in charge, sometimes because of the difficulties inherent in their location, in a few cases perhaps because of dishonesty. The officers in charge were subjected to many discouraging difficulties: irregular and insufficient supplies of materials and funds; frequent interference—by conscription, enrolling officers, and congressional inquiries—with their male operatives, foremen, agents, and officers; lack of proper direction from the quartermaster general's office; and the general confusion and demoralization in the service in districts remote from Richmond.

The proper equipment of the shops was retarded by the great scarcity not only of the requisite machinery but also of the means of making it. By an act of October 8, 1862, the congress authorized the president to import free of duty machinery or materials for the manufacture of clothing and shoes for the army. The privileges of the act might be extended to companies or individuals under regulations prescribed by the president.[309] Under this act some machinery was brought in for the government; but that a much larger quantity was imported for the privately owned cloth and yarn mills is clearly indicated by their rapid increase, although this development was aided very much by the state governments, notably by that of North Carolina.

The army had been supplied with clothing during the first year of the war by various means, more or less fortuitous. The quartermaster general had come to rely for his supply of cloth upon the private cotton and woolen mills, all of which were small. By the summer of 1862 he found it necessary to enlarge his contracts with the factories. The factory owners had just begun to realize what a boon a war may be to business men. Before secession their establishments had led a precarious existence because of the stifling competition of the great northern and English mills.[310] The war had cut off that competition, there was a strong demand for more than they could turn out, and prices were rising rapidly. But this behavior of prices deranged all estimates of the purchasing officers, ate rapidly into the appropriations, and was the subject of constant complaint on the part of the quartermaster general. The underlying causes of this phenomenon are so familiar that they need not be detailed here. A contributing cause was the faulty purchasing organization which resulted in frequent

---

[309] "An act to encourage the manufacture of clothing and shoes for the army," *ibid.*, chapter 31.

[310] See William Bailey to Governor John Milton, Tallahassee, June 15, 1864, in *War of the rebellion: a compilation of the official records of the union and confederate armies* (Washington, 1880-1901), series 4, vol. 3, p. 500.

competition between representatives of the different depots. Another, and very obvious one, was that the general stock of cloth on hand in the south in the spring of 1861 was about exhausted. Although competitive bidding between government agents was corrected to some extent by confining the contracts of each depot to a specified territory,[311] competition by the general public was not eliminated. The manufacturers in many cases were finding larger profits in the rising public markets than in contracts with the quartermasters, and either hesitated or declined outright to bind themselves to long term agreements at fixed prices. Those who refused to contract were sometimes threatened with the impressment of their goods; but this mode of procuring supplies was distasteful to the higher authorities in these early days of the war and it does not appear that impressment of cloth was resorted to at this time. Many of the factory owners, to be sure, willingly continued to sell to the government a large portion of their products. For the others a means of coercion was preparing.

The first confederate conscription act, approved April 16, 1862, made no provision for the exemption of any person embraced by its terms; but it was followed five days later by an act which exempted directly several classes of persons and in addition "superintendents and operatives in wool and cotton factories, who may be exempted by the Secretary of War."[312] Although the

---

[311] A. C. Myers to M. G. Harmon, December 30, 1861, to V. K. Stevenson, January 4, 1862. Letter books, quartermaster general's office, in the confederate archives, United States war department, Washington, 3: 50, 76.

[312] "An act to exempt certain persons, etc.," *Confederate statute. at large*, 1 congress, 1 session, chapter 74.

discretionary power conferred upon the secretary by this act clearly gave him the means of bringing pressure to bear upon the manufacturers to force them to contract with the government, exemptions seem to have been allowed so liberally that the situation was altered but little.[313]

The second conscription act, September 27, 1862, extended the military age to include men from thirty-five to forty-five years of age. It was followed by the second exemption act, October 11, 1862, which repealed the previous one and substituted a much more detailed list of persons to be exempted under certain conditions. Among these were shoemakers, tanners, wagonmakers, and other mechanics, and "superintendents and operators in wool and cotton factories, paper mills and superintendents and managers of wool carding machines, who may be exempted by the Secretary of War; Provided, the profits of such establishments shall not exceed seventy-five per centum upon the cost of production, to be determined upon oath of the parties"—subject, upon proof of violation to the secretary of war, to cancellation of the exemptions and the refusal of any whatsoever in the future.[314]

The provision which restricted profits derived from the labor of exempted persons to seventy-five per cent, on condition of the grant of exemption, was plainly an effort

---

[313] There is some evidence that the conscription act for a time reduced factory production and embarrassed the quartermaster general by checking the flow of supplies to the army. See Myers to G. W. Randolph, May 23, 1862, in *Rebellion records*, series 4, vol. 1, p. 1127.

[314] *Confederate statutes at large*, 1 congress, 2 session, chapter 45.

to check extortion and profiteering rather than to assist the quartermaster general, for it applied to goods sold to civilians as well as those sold to the army. It is significant that such a large profit was conceded. The southern people had always opposed in principle any direct interference by the government with legitimate private business, but feeling against the greedy extortioner who was blamed for the high prices and consequent privations, was so intense that all theoretical considerations were lost sight of The exemption provisions were put into effect by general orders from the war department in November, 1862, which provided that each application for exemption should be accompanied by affidavit and be returned to the local commandant of conscription, and that the exemptions should be granted or refused by the enrolling officer.[315] This part of the law remained in effect until February, 1864. As far as providing for the labor needs of the factories was concerned, it seems to have worked fairly well for a while. Late in 1863 the pressure for men for the army made it harder to convince the enrolling officers of the necessity for the exemptions. The factories were so dependent upon this labor that they rarely dared to refuse to make contracts with some department of the government. When a disagreement arose the threat to enroll their "exempts" was usually enough to bring them to terms.[316] It came to be felt that the profit allowed was too high, especially in view of the rapid turnover and of the

---

[315] General orders no. 82, adjutant, and inspector general's office, November 3, 1862; 8, "Provisions against extortion," in *Rebellion records*, series 4, vol. 2, pp. 160-168.

[316] A. C. Myers to W. E. Jackson, March 26, 1863, Letter books, quartermaster general's office, 6: 172.

hardships and misery of the people. Again, the system worked badly because there was no incentive to the manufacturer to keep down costs; on the contrary he was tempted to increase them. There was much suspicion that cost accounts were padded; but there was as yet no adequate method of inspecting them, and the penalties were rarely imposed.

These considerations may have had some weight in the passage of the conscription act of February 17, 1864, but the main reason for that stringent law was the stern necessity of filling the ranks of the depleted army. By this measure the conscription age limits were extended to seventeen and fifty years, while all exemptions, except for religious opinions, were repealed, and a new and shorter list was substituted. Workers in factories were no longer exempted, but the president might grant details either from persons between forty-five and fifty years of age who had been enrolled but not yet called into the ranks, or from the army in the field.[317] Under the general orders which put this law into effect the granting of the details was virtually placed in the hands of the local enrolling officer, who could make the grant for no longer than sixty days. At the end of that time the contractor must make a new application.[318]

Nothing was said in this act about the seventy-five per cent profit, and if it had not already ceased to be regarded as a proper one, it soon afterwards was

---

[317] *Confederate statutes at large*, 1 congress, 4 session, chapter 45, sections 10, 11.

[318] "General orders no. 26," adjutant general's office, March 1, 1864, in *Rebellion records*, series 4, vol. 3, pp. 178-183; also Circular no. 8, bureau of conscription, March 18, 1864, ibid., 217-224.

abandoned. Perhaps General Lawton, who became quartermaster general in August, 1863, was a keener business man than Myers; at any rate, although the cost plus plan was continued, the profits allowed in 1864 were usually about thirty-three and one-third per cent, and sometimes they ran as low as twenty-five per cent."[319] It was usually stipulated in the new contracts that the factory should sell at least two-thirds of its cotton goods and all its woolens to the quartermaster's bureau; and that the quartermaster general in turn should endeavor to procure from the secretary of war enough details to provide the necessary labor for the factory. It does not appear that the manufacturer was placed under any restriction as to the prices he should charge in the public market. Lawton gradually adopted a rigorous policy with his contractors and insisted upon the right to examine the books of any concern whose cost estimates were suspicious.[320]

In the spring of 1864 the affairs of the bureau were in such confusion in the gulf states that General Lawton put the whole business of contracting with the cloth factories of that region into the hands of Major Cunningham, who

---

[319] W. G. Ferguson to A. F. Cone, April 13, 1864, "Letters received," quartermaster general 's office; A. R. Lawton to G. W. Cunningham, July 9, 1864, to W. G. Ferguson, September 12, 1864, Letter books, quartermaster general's office, 8: 341, 9 : 96.

[320] Report on cost of production at Logan factory, Richmond, July 1, 1864, "Citizen's vouchers," quartermaster general's office; Lawton to Whitehead and Broocks (proprietors of above factory), July -, 1864, *ibid.*; Lawton to Cone, August 19, 1864, Letter books, quartermaster general's office, 9: 2. This last is with reference to an investigation of the Manchester (Virginia) and Petersburg factories.

had distinguished himself by his efficient conduct of the depot and shops of Atlanta.[321] Within two months Cunningham was able to report a considerable increase in government receipts from the factories in Georgia, Alabama, and South Carolina. Some factories had refused to make contracts with him on the ground that they had contracts, though small ones, with other bureaus or departments. "In such exceptional cases," he wrote, "I am much embarrassed for want of proper orders by which to control the detail of their men, as they have but to contract with some other Department and I am without remedy."[322] This was already a grievance with Lawton. The bureau of subsistence, the bureau of ordnance, and the navy department all had use for cloth of one sort or another; and each had obtained factory contracts. The contracts were usually small, but the factories by means of them had obtained the coveted and necessary details, and refused to make contracts with Lawton. They could dispose of their surplus product to better advantage than to the government. Lawton thereupon tried to induce the other bureaus and the secretary of the navy to transfer their contracts to him on condition that they receive the same amount of cloth as before. By getting control of the details, he could exact a larger part of the factory output. They agreed, but Colonel Northrop, the commissary general, insisted upon so much more cloth than Lawton could promise that the transfer was never completed.[323]

---

[321] Lawton to Cunningham, April 9, 1864, *ibid.*, 8: 186; also circular issued by Lawton, April 23, 1864, a copy of which is in "Personal papers" of Cunningham, confederate archives, United States war department.

[322] Cunningham to Lawton, May 12, 1864, *ibid.*

[323] Lawton to S. P. Moore, March 31, 1864, to L. B. Northrop,

The secretary of the navy, however, consented to turn over to the quartermaster general his contract with the very important Vaucluse factory in South Carolina, which had been selling to the navy department only one tenth of its goods. It had refused to contract with the army because "independent in the matter of details." The transfer, said Lawton, would secure to the navy its regular supply of material, and "will enable this Bureau to increase the production at Government account by subjecting all Factories to a uniform system."[324]

The bureau of conscription, which by order of the secretary of war had been charged with the granting of details for factory labor, insisted upon a very stringent supervision of the business of the contractors. Doubtless this was because the conscription officers wished to make a good showing in keeping men in the army, but the effect was to strengthen control over the factories. In one case, which seems to have been typical, the enrolling officer refused a petition for a detail to a cotton factory in South Carolina unless the proprietors would agree to sell their surplus to the people at the same rate as to the government. The superintendent of conscription on appeal sustained the enrolling officer and explained that in order to stimulate production the earlier supervision

---

to J. Gorgas, to A. L. R.ives, April 1, 1864, to Cunningham, June 4, September 9, 1864. Letter books, quartermaster general 's office, 8: 139-141, 269, 9: 163. Lawton from Gorgas, April 4, 1864, from John de Bree, April 16, 1864, from Northrop, April —, 1864. "Letters received," quartermaster general's office.

[324] Lawton to Stephen R. Mallory, September 3, 1864, to Cunningham, September 29, 1864. Letter books, quartermaster general's office, 9: 53, 163.

had not imposed restrictions upon prices to the public; but now that production was probably at its maximum, it was proposed "to extend the supervision to the rates charged and to insist as a condition to the detail upon the disposal at reasonable rates of the articles manufactured." The thirty-three and one third per cent profit allowed by the quartermaster's bureau would be adopted as the rule. "The books of the concern should at all times be open to the inspection of the Enrolling Officer," and limiting the details to specific periods would result in frequent inspections and guard against the abuses of the cost plus plan.[325]

The conscription bureau was abolished by law in March, 1865, and its duties devolved upon the general officers commanding the reserves in each state. But the confederacy was in dissolution, and in the confusion of the break-up which followed there is little likelihood that any new policy towards the factories was instituted.

Through the virtual control of raw materials the quartermaster's bureau obtained another line of control over the textile factories. Raw cotton was plentiful in the southern tier of states until the end of the war; but after the quartermaster general had established a close supervision over the traffic of the railways and had put into force a system of priorities for army supplies, such as food and munitions, the feeble railways were so clogged that there was little opportunity for factories to get cotton, except locally, unless the quartermasters consented to its

[325] C. B. Duffield to the secretary of war, January 16, 1865, "Letters received," quartermaster general's office. This is an endorsement on the petition of W. A. Pinger to be detailed as superintendent of Valley Palls cotton factory. The first endorsement is by Lieutenant G. N. Marshall, enrolling officer.

transportation. The Virginia factories came to this situation by the end of 1863; in 1864 even those in the cotton states were dependent upon the quartermaster general for a large portion of their supplies. When the raw material must be brought, under military orders, over an army-controlled railroad to a factory operating with labor detailed from the army, there was no way for the manufacturer to avoid selling to the government as much of his product as was demanded.[326]

Wool became scarce at the very beginning of the war. In fact, the south produced but a small part of the wool it consumed, and within a little more than a year some of the most productive regions were in the hands. of the union armies. Agents were sent out early in 1862 into all the wool growing sections under orders to purchase all that was on the market at reasonable prices and to impress the rest. Great expectations of the Texas crop were cherished, but the clip of 1862 was sold before the government agents were in the field, and the major part of the clip of 1863 did not reach the Mississippi river in time to cross before the taking of Vicksburg and Port Hudson. The tax in kind, from which much had been expected, yielded little. Importations, through the blockade, from England, though helpful and important, still left a serious shortage. In 1864 a large proportion of the trousers sent to the army were entirely of cotton. In 1863 the government bureaus had obtained a virtual monopoly of the wool on the market—some escaped by

---

[326] Lawton to W. G. Ferguson, September 12, 1864, Letter books, quartermaster general's office 9: 96. For a discussion of the confederate railway problem, see the *American historical review*, 22: 794-810.

clandestine exchange—and were able to require the woolen factories to make up the whole of it for government account. This monopoly was maintained for the rest of the war.

Thus far nothing has been said of the relations of the several state governments to the factories and of the consequences to the business of the confederate officials. North Carolina was the only state whose policies in this respect will be noted here. In the early days of the war when commutation was allowed the soldier who provided himself with clothing, the general assembly adopted a resolution that the state should furnish its soldiers with clothing, shoes, and blankets and receive in payment the commutation money allowed by the war department.[327] This was agreed to by the war department, and, though it seems to have been unsatisfactory to some of the North Carolina troops in the field and proved a source of confusion in the quartermaster's bureau, the plan continued in force even after the abolition of the commutation in October, 1862.[328] Thereafter the state sold the clothing to the confederate quartermasters on the condition that it was to be issued to North Carolina troops exclusively. The state, of course, obtained these supplies either by contract with the factories within its borders, or by purchase abroad. "Exempts" and details were evidently allowed to these factories by the war department as well as to those contracting with the quartermaster general.

---

[327] *Public laws of the state of North Carolina Passed by the general assembly at its extra session, 1861*, pp. 129-130. The resolution was adopted September 20, 1861.

[328] Myers to J. G. Martin, adjutant and inspector general of North Carolina, May 22, June 5, 1862, Letter books, quartermaster general's office, 4: 230, 275.

General Myers, who had previously endeavored without success to persuade the North Carolina authorities to cancel the agreement, took the position that the abolition of commutation had itself abrogated the arrangement with the state. He therefore renewed his appeals that North Carolina place herself on the same basis as other states by surrendering to the quartermaster's bureau her contracts with the cloth factories and tanneries—from which the confederate officials were virtually excluded—thus allowing troops from other states less fortunately supplied with such factories to share their products. The state authorities—Vance was now governor—steadily refused.[329]

This situation lasted for nearly two years. Early in 1864 when the clothing situation was desperate, Quartermaster General Lawton instructed the post quartermaster at Raleigh to endeavor to get contracts for the bureau with the factories of North Carolina, if it could be done without coming into conflict with the state authorities.[330] This quartermaster, Major Peirce, first tried again to induce the state authorities to agree to allow their contracts to be taken over by the confederacy in order that troops from other states might share in the clothing. Again the reply, by direction of Vance, was flat refusal to allow any of the North Carolina clothing to go to other troops until the governor should first satisfy himself

---

[329] Myers to Governor H. T. Clark, June 12, 1862, Letter books, quartermaster general's office, 4: 292, to Zebulon B. Vance, September 17, December 8, 1862, Vance's letter books, 1: 8, 68; Vance to Myers, October 14, 1862, *ibid.*, 25.
[330] Lawton to W. W. Peirce, February 25, 1864, Letter books, quartermaster general's office, 8: 50.

that there was a surplus.[331] Lawton next inquired whether the state would object to the quartermaster's bureau making contracts on its own account with the North Carolina factories. He was informed that, owing to the scarcity of wool, the product of the woolen mills would not be shared, but that the confederate government was at liberty to make contracts with the cotton factories, since but few furnished the state more than one third of their output.[332]

Lee's men were ragged and barefoot; the needs of the army would admit of no further dalliance. Lawton now determined to move directly upon the factories themselves. More than one third of the whole number of textile factories within the entire confederacy east of the Mississippi were in North Carolina, and among these were also the largest and best equipped.[333] Though exact figures are lacking, it seems probable that the factories of North Carolina from which the confederate government hitherto had been in effect excluded, were producing half of all the cloth in the confederacy in 1864.

In August, 1864, Lawton gave instructions that

---

331 Peirce to Lawton, with enclosures, March 17, 1864, "Letters received," quartermaster general's office.

332 Peirce to Lawton, with enclosures, April 2, 1864; Gatlin to Peirce, April 14, 1864. "Letters received," quartermaster general's office.

333 See the statement of Lawton to .James A. Seddon, September 28, 1864, in *Rebellion records*, series 4, vol. 3, p. 691. There are lists of the factories in the report of Major .James Sloan to W. B. B. Cross of the quartermaster general's office, September 1, 1864, in "Letters received," quartermaster general's office, and in "Cash book," quartermaster general's department of North Carolina, in the archives of the North Carolina historical commission, in Raleigh, p. 1.

careful investigation be made of the prices at which cotton and woolen goods, including yarns, could be had in North Carolina; also that a list of all the factories be made, with a report on the capacity of each, what part of its product was contracted to the state, and whether it would make a contract with the confederate government.[334] The factories seemed reluctant to send in written statements, and an agent had to visit them. It developed that although the state received only about one third of the product on the average, the factory owners were unwilling to contract any more of it to the confederate authorities, alleging that what was left was needed to barter for raw materials and labor, as the currency was valueless.[335] Just at the time the agent of the quartermaster general was making the round of the factories and hearing their several reasons for their inability to enter into contracts, the bureau of conscription began calling in the details whose services the factories had been enjoying.

"This," wrote Lawton's agent, "looks to the factories as if everything was arranged in Richmond with special reference to forcing them into measures, and they so express themselves. It is not very material whether they are correct or not; but if they think so it makes it easier

---

[334] Lawton to Major L. R. Chisman, August 2, 1864, Letter books, quartermaster general's office, 8: 385.

[335] Circular of Chisman to factory owners, August 15, 1864; Major Sloan to Cross, August 24, September 1, 1864. "Letters received," quartermaster general's office. See also Lawton to Sloan, August 19, 1864, to William A. Miller, September 2, 28, 1864, in Letter books, quartermaster general's office, 9: 2, 48, 158; and reports of Miller to Lawton, September 12, 17, 1864, in "Letters received," quartermaster general's office.

for me to make contracts."[336] Although Lawton almost certainly had nothing to do with the action of the conscript bureau, which in fact was calling in details in other states as well and interfering seriously with his activities, Governor Vance shared the suspicions of the factories, his wrath blazed up, and he seized the opportunity for exercising his favorite recreation of "trying a tilt with the Confederate government." He charged heatedly that Lawton was trying to break up the state's business with the factories in order to seize it all for his own department, and he demanded impetuously to know whether the confederate administration was giving countenance to the nefarious scheme. Lawton, with some asperity, denied the charge, and recounted the history of the troubles over the North Carolina factories, the growing scarcity of supplies available for the troops of other states, and the injustice and ill-consequences of the selfish policy followed by the state authorities.[337]

Unfortunately the records thus far available do not disclose much concerning the consequences of this controversy. That the state continued to send large amounts of shoes, clothing, and blankets to its troops in the field is clear from examination of its account with the confederate government, month by month, from September 30, 1864, to January 31, 1865.[338] That the

---

[336] Miller to Lawton, September 19, 1864, *ibid.*

[337] Vance to Seddon, September 19, 1864, in *Rebellion records,* series 4, vol. 3, p. 671; Lawton to Seddon, September 28, 1864, *ibid.*, 690-692. See also Vance to General Holmes, October 25, 1864, *ibid.*, 746.

[338] For September, 1864, $674,094; for October, $562,722.50; for November, $467,133; for December, $852,844; for January, 1865, $456,518. Manuscripts in the archives of the

stores of shoes and clothing in the quartermaster's depots in Virginia and the Carolinas increased considerably is indicated by frequent statements of Lawton in December, 1864, and January, 1865, that plenty was on hand.[339] As it is unlikely that large importations could have come into "Wilmington, where the blockade was tightening rapidly, one might infer that the North Carolina factories had supplied the increase, were it not for a tabulated report in February, 1865, by the bureau of conscription of persons detailed for labor in the various states. North Carolina shows a total of 261, against 387 for South Carolina, 797 for Georgia, and 927 for Virginia. There were no details shown for contractors with the quartermaster's bureau in North Carolina.[340]

The experiences of the government bureaus in procuring leather call for a separate account. By far the greatest consumer of leather was the quartermaster's

---

North Carolina historical commission. Even when allowance is made for higher prices, these figures show much larger sales than for any previous period of equal length, if it may be assumed that the earlier accounts are in full.

[339] Lawton to Lieutenant Colonel J. L. Corley, December 12, 1864, to A. B. Bragg, January 3, 1865, in *Rebellion records*, series 1, vol. 42, part 3, p. 1268, and vol. 46, part 2, p. 1009; to Miller and others, committee of the house of representatives, December 12, 1864, Letter books, quartermaster general's office, 9: 345; to Miller, January 27, 1865, in *Rebellion records*, series 4, vol. 3, pp. 1039-1041. Against Lawton's optimistic statements in the last citation above may be set one from Cunningham to Lawton, March 17, 1865, deploring the scarcity of cotton goods for shirts and underwear, his inability to pay contractors, and the suffering in the armies for each of these articles. "Miscellaneous letters received," quartermaster general's office.

[340] *Rebellion records*, series 4, vol. 3, pp. 1099-1109.

bureau, which furnished shoes for the army and harness for baggage trains. The ordnance bureau also used leather for artillery harness and saddles. It was not until the early spring of 1862 that leather became noticeably scarce. In June, post quartermasters were instructed to make a special effort to procure shoes either by purchase or by having them made by workmen under their own supervision.[341] In July the supply was so short that experiments were made with canvas topped shoes, and not infrequently afterwards these cloth shoes were used. The tanneries everywhere reported a shortage in the supply of hides and raised their prices. This resulted in an order of the secretary of war fixing prices for the several grades of leather and authorizing impressment when these prices were not accepted.[342] The commissary officers who killed beeves for the army were directed to turn over all hides to the quartermasters. Although agents were sent out in search for them, only a small proportion of the hides which the country afforded was obtained. The indifference and carelessness of the commissary officers, who allowed the green hides to spoil for lack of a little attention, the unwillingness of farmers to give up their only chance to obtain leather for their own use, and the preferenee of the tanners for the greater profit in private traffic combined to defeat the expectations of the

[341] Myers to Captain F. W. Dillard, June 11, 1862, to Madison McAfee, Roland Rhett, and others, June 12, 1862, to Major T. F. Washington, June 19, 1862, to Major J. L. Calhoun, June 23, 1862. Letter books, quartermaster general's office, 4: 290, 294, 320, 329.

[342] W. S. Downer and J. B. Ferguson to Larkin Smith, July 28, 1862, "Letters received," quartermaster general's office, and the orders endorsed thereon by the secretary of war.

quartermaster general.[343]

As in the case of the cloth factories, the quartermasters began to press the advantage given them by the conscription acts. Tanners who found themselves or their hands liable to conscription were glad to make contracts which would enable them to stay out of the army and keep their business going. The contract usually provided that the tanner should receive one third of the leather made from the hides furnished by the government. He was usually required to furnish bond to cover the value of the hides, and, frequently, to sell to the government his third of the leather at a fixed price.[344] The post quartermaster of each larger district was given exclusive control over contracts within his district.

As soon as the bureau began to obtain leather it set up its own shops for the making of shoes. By an act of congress of October 9, 1862, the president was authorized, on requisition of the quartermaster general, to detail from the army persons skilled in the manufacture of shoes, not to exceed two thousand; and it was made the duty of the quartermaster general to place them at suitable points in shops, under regulations prescribed by himself, to make shoes for the army. These

---

[343] "General orders no. 101," adjutant general's office, December 9, 1862, in Rebellion records, series 4, vol. 2, p. 232. See also Smith to Dillard, July 18, 1862, Letter books, quartermaster general's office, 4: 400; J. J. Smylie to R. P. Valler, October 17, 1862, "Letters received," quartermaster general's office.

[344] A contract, a printed form filled in, made by J. M. Hackett and A. B. Smith, High Point, North Carolina, with Major Waller, quartermaster at Richmond, November 1, 1862, may be found, *ibid.*

details were allowed extra pay.[345] The most conspicuous and successful of these government shoe shops was that at Columbus, Georgia, under charge of Major F. W. Dillard. Others of importance were at Richmond, at Atlanta, and somewhat later at Montgomery. These government shops had tanneries attached, but received a good part of their leather from private tanneries under contract. In some parts of the country, as in the Carolinas, private shoe factories continued in operation for a time and but few government shops were established. Due in part to the greater economy effected by concentration, a consolidation of the government shoe factories began in the summer of 1863,[346] and was carried further in 1864.

The visible supply of leather, and therefore of shoes, harness, and saddles, was declining rapidly.[347] Although

---

[345] *Confederate statutes at large,* 1 congress, 2 session, chapter 37. See also W. W. Lester and William J. Bromwell, *Digest of the military and naval laws of the confederate states of America* (Columbia, S. C., 1864), 93.

[346] For instance the Atlanta shops were transferred to Columbus and there consolidated with Dillard's; the Danville workers were sent to Knoxville. See Myers to Cunningham, April 25, 1863, to W. G. Sutherlin, May 13, 1863, in Letter books, quartermaster general's office, 6: 316, 379. Several shoe shops at other points in Alabama were consolidated, first at Selma, later at Montgomery. Inspection report, Lieutenant Colonel E. E. McLean to A. R. Lawton, December 6, 1863, in "Personal papers" of McLean, in the confederate archives; A. R. Lawton to Gillaspie, June 23, 1864, Letter books, quartermaster general's office, 8: 305.

[347] During the battle of Gettysburg, General Lee's quartermaster telegraphed a requisition for 20,000 pairs of shoes. Myers' reply is significant: "I have been in anticipation of receiving stores from you and not to supply you." Myers to Corley, July 3, 1863, Letter books, quartermaster general's office, 6: 539.

the blockade runners continued to bring in both leather and shoes, so many cargoes were destroyed that the relief furnished was wholly inadequate.[348] The loss of the Mississippi river sharply reduced the number of beef cattle that had been crossing from Texas and western Louisiana. The best cattle regions east of the Mississippi were now overrun by the enemy or were being subjected to destructive raids. There can be no doubt that the number of cattle within the confederate lines east of the Mississippi fell off rapidly. A situation which was bad enough was made worse by the frequent gross carelessness or inefficiency of the officers and agents charged with preserving, collecting and shipping the hides, by the refusal of commissaries, who desired to make purchases with hides, to turn them over to the quartermasters,[349] and, as demoralization grew, by the absence of proper control over distant quartermasters and agents, and by downright theft. Thousands of hides were kept out of the hands of the government by speculators, and evidence is not lacking that some minor officials themselves carried on private speculation in this much sought after commodity. The system of collecting hides had become so demoralized that in August, 1864, Major Dillard, whose success with his shoe factory at Columbus has already been mentioned, was given

---

[348] Lawton to General Robert E. Lee, January 20, 1864, to General Joseph E. Johnston, February 9, 1864. *Ibid.*, 7: 554, 8: 15.

[349] There are numerous instances of this. Dillard from J. F. Cummings, November 16, 1864, from .T. L. Locke, November 28, 1864, A. Myers to Goodman, December 27, 1864, Goodman to Lawton, January 9, 1865. "Letters received," quartermaster general's office.

supreme control over the business in all the region south of North Carolina and Tennessee and over contracts for tanning as well.[350]

The arrangements for tanning were also in need of revision. As the supply of hides fell off, fewer tanneries were needed. The constant pressure for the return of details to the army, and the difficulty of safeguarding the interests of the government in out-of-the-way places made it clearly advisable to reduce the number of government contractors and return the unnecessary details.[351] Under instructions from the quartermaster general the number of contract-tanneries in Virginia was reduced in October, 1864, from one hundred and twenty-five to about a dozen. In South Carolina, Georgia, and Alabama, Major Dillard reported a total reduction of from two hundred to sixty-seven, which was thought still too many.[352] The tanneries which were retained were required to work entirely for the government—since the quartermaster's bureau was taking all the hides whether on the market or in private storage—and to tan at a fixed money rate instead of taking pay in leather.

The business of making shoes for the army was now monopolized by the government shops through their

---

[350] Circular of Quartermaster General Lawton, September 19, 1864, personal papers of F. W. Dillard, in the confederate archives.

[351] "The facilities for tanning are so ample throughout the whole country that there would be no difficulty in tanning for money and on reasonable rates ten times the number of hides that can be obtained." Lawton to Major J. C. Green, August 24, 1864, Letter books, quartermaster general's office, 9: 13.

[352] Lawton to W. G. Ferguson, October 12, 1864, to Dillard, October 21, 1864.

control of hides and leather and their lower cost of production.[353] Moreover, they could easily make up all of the scanty supply of leather available. But even the government's own establishments were not free from business worries. The uncertain supply of raw materials was perplexing; but the interference of the conscription bureau with their details of workmen was even more annoying. The records of the quartermaster general's office show graphically how closely the service was combed for men to fill the ranks in the field. Every list of workers or officers was scrutinized with care lest some able-bodied man should slip through. The complaint of Captain W. M. Gillaspie, in charge of the shoe factory at Montgomery, must have described the experiences of others in similar positions. "It takes fully one half of the time of one of my clerks and about as much of my own to attend to the detailing of men. I no sooner have a man trained and somewhat efficient than he is ordered to report to camp, and then after considerable delay I have his services replaced by a new one and an invalid, in most cases, who is one half the time in the Hospital and the other half not able to do much work."[354]

So scanty are the records of their activities during the last months of the war that little can be said of the condition of the factories or of their relations with the government. Some were destroyed by Sherman and other raiders; some closed down for want of material, or labor, or funds; some continued to the end. Though no change

---

[353] A. U. Davis to Governor M. L. Bonham, July 25, 1863, *ibid.*
[354] Gillaspie to Cross, assistant quartermaster general, November 4, 1864, *ibid.* The Richmond shoe factory was required to give up to the army 66 out of 171 employees in January, 1865

in the general policy of the government is discoverable, there is some indication that in the weakness and confusion which preceded dissolution, the system of control, once so rigid, relaxed.

It seems significant that the confederate government never devised nor even attempted to devise any civil machinery for the control or regulation of the factories upon whose production so much depended. The reason is clear: the supply of the army was regarded as solely a military problem. It was an army affair; leave it to the war department! Though an urgent, it was a temporary problem which would be over with the coming of peace. Perhaps the southerners disliked the idea of a permanent arrangement for interfering with private property rights; but probably they never thought of that at all. The military control which was gradually established was rigid, but partial, uneven, short-sighted. The sole purpose was to get supplies for the army at as reasonable a cost as possible—to exploit the factories, not to develop them for the benefit of the community at large. After all, these things were inevitable under a government struggling for bare existence and dependent upon undeveloped and inadequate resources.

Charles W. Ramsdell

University of Texas
Austin

Part Four

Ramsdell's Book Reviews Are Works of Art

Compiler's Note: *There is an unbelievable amount of history in each of the fifteen book reviews in this section. Ramsdell had to include a lot in order to provide background and put the subject in perspective. Ramsdell pulls no punches. He reviewed way more than fifteen books in his career, but in this fifteen are outstanding, well-known books by distinguished historians.*

List of books reviewed

R. E. Lee: A Biography, 4 vols.,
by Douglas Southall Freeman

The Civil War and Reconstruction,
by J. G. Randall

The Repressible Conflict, 1830-1861,
by Avery Craven

The American Civil War: An Interpretation,
by Carl Russell Fish

State Rights in the Confederacy,
by Frank Lawrence Owsley

Life and Labor in the Old South,
by Ulrich Bonnell Phillips

The Story of the Confederacy,
by Robert Selph Henry

Constitutional Problems Under Lincoln,
by James G. Randall

Guide to the Study and reading of American History, by Edward Channing,
Albert Bushnell Hart, and Frederick J. Turner

Bedford Forrest, The Confederacy's Greatest Cavalryman, by Eric William Sheppard

Southern Editorials on Secession,
compiled by Dwight Lowell Dumond

The Secession Movement, 1860-1861,
by Dwight Lowell Dumond

Aeronautics in the Union and Confederate Armies, With a Survey of Military Aeronautics Prior to 1861. Vol. 1, by F. Stansbury Haydon

Lincoln Takes Command,
by John Shipley Tilley

John Brown, Soldier of Fortune: A Critique,
by Hill Publes Wilson

R. E. Lee: A Biography.
By Douglas Southall Freeman.
4 volumes. (New York and London: Charles Scribner's Sons, 1934, 1935. Pp. xiv, 647; xiii, 621: xiii, 559: ix, 594. $15.00.)[355]

As nearly as any work may, these four volumes constitute the definitive life of Robert E. Lee; for, while even the indefatigable researches of Dr. Freeman over nearly twenty years cannot possibly have brought to light every scrap of evidence, it is improbable that any thing yet to be discovered will materially change the story he has told or seriously impair the judgments he has formed on Lee's character, actions, and career. Although the original plan was for only one volume, as the material in his hands accumulated Dr. Freeman wisely chose to give himself full scope. The result is a full, clear narrative that moves with dignity, without hurry or prolixity, and frequently with eloquence. Of the more than 2300 pages, about 450 are given to Lee's life prior to his resignation from the United States army in April, 1861 (of which 100 cover his participation in the War with Mexico), a little more than 1500 to his services to Virginia and the Confederacy, and about 300 to the presidency of Washington College. More important even than the discovery of new material—or at

[355] This is the review exactly as it appeared in *The Journal of Southern History,* Vol. 1, No. 2 (May, 1935), 230-236.

any rate more interesting to this reviewer—is the careful analysis and weighing of the sources, especially when they are conflicting, the explanation of the elements which went to the formation of Lee's character and habits, the description of his steady growth in professional competence, and the exposition of the methods by which he solved his military problems.

How much of his high qualities of mind and character was derived from the ancestral Lees and Carters, how much came of the severe lessons inculcated in childhood or from an enlightened self-discipline no one can say with confidence. Certainly his forebears were men and women of character, but the reader gets the impression that innate honesty, simplicity of soul joined to the courtesy and kindliness of a true gentleman, and the precise workings of a high order of intelligence are the best explanations of both his military successes and the hold he acquired over the affections of all southerners and, eventually, of discerning northerners.

When at the age of thirty-nine Lee got his first experience of warfare in Mexico he had seen seventeen long years of service in the Bureau of Engineers and had reached no higher rank than a captaincy. His experiences in Mexico were to reveal his abilities and to teach him many things. Freeman suggests that Lee was inspired by General Scott's example to audacity, that he learned from him the value of a trained staff in the development of strategical plans, the importance of careful reconnaissance, of field fortifications, of the great possibilities of flank movements, the relations of communications to strategy, and that "Lee concluded, from Scott's example, that the function of the commanding general is to plan the general operation, to acquaint his corps commanders with that plan, and to see that their troops are brought to the scene of action at the proper time; but that it is not the function of the commanding general to fight the battle in detail. . . . Whether he was right in this conclusion is one of the moot questions of his career." He had no opportunity to study the use of cavalry and had to learn that in 1862. Nor did he, in an army of only 10,000 men, have a chance to observe large scale operations or transportation by railroad. Between 1848 and 1861 he was able to advance his military training only during the three years while he was superintendent at West Point by the study of Napoleon.

It is well known that Lee was opposed to secession and that his resignation from the army in April, 1861, was based only upon what he felt was due to his state and his people. His high reputation was known to the authorities of Virginia and caused him to be made commander of the military and naval forces of the state. The value of his services in mobilizing the Virginia volunteers and in selecting points of defense has been obscured by the fame of his later campaigns. so that not the least of Freeman's distinctive contributions is his account of Lee's work as a military organizer and administrator in the early summer of 1861.

When the Confederate government took over control of the Virginia volunteers, Lee, who had been raised to the rank of general in the Confederate army by Jefferson Davis, remained in Richmond until one week after the battle of Manassas. Then he went into the mountains of western Virginia to begin his first independent campaign.

Here again Dr. Freeman has given us a clear account of what has hitherto been much confused. Lee faced immense difficulties. He was sent out to coordinate, not to command, the scattered forces, although he did later take over command. But the principal officers gave him infinite trouble with their mutual jealousies and bickerings. It rained incessantly; the roads were quagmires of "unfathomable mud"; food and forage were inadequate; the men were weakened by measles and other sickness. When on two occasions he worked out plans of attack against the Federals, he was frustrated partly by the rains but more by the incompetence and quarrels of his officers as well as by his own unwillingness to be peremptory with them. Lee returned to Richmond late in October without recovering western Virginia, the public confidence in him virtually gone. It is to the credit of Jefferson Davis that he understood Lee's difficulties and stood by him. Lee, for his part, had learned much in the mountains.

In exactly one week after his return he was sent to command the South Carolina-Georgia coast where the Union navy was threatening. His work there, largely that of an engineer, was so nearly perfect that the Confederates were able to hold the defenses he laid out until Sherman's army took them in the rear in 1865. He was called back to Richmond early in March, 1862, to serve as military adviser to President Davis but without real authority. It was an uncongenial task, but he set to work. He was chiefly responsible for the resort to conscription, but his plan was badly mangled in the legislation by Congress. After Joseph E. Johnston had retired from the Manassas front to face McClellan on the

Peninsula, Lee was able to suggest the plan for the brilliant campaign by which "Stonewall" Jackson frightened Washington and prevented the Federal forces in northern Virginia from going to the aid of McClellan—a far-reaching strategic plan which was nearly wrecked by Johnston whose ideas for the defense of Richmond never went beyond the concentration of all available Confederate forces in front of that city and who never quite grasped the daring conceptions of Lee. But Lee's part in the movement was unknown both to the public and the army, and when Davis placed him in command of the army on June 1, after Johnston was wounded, he had never actually conducted a battle and his reputation was still clouded.

It is manifestly impossible, within the limits of this review, to trace Freeman's account of each of Lee's campaigns; but something should be said of his method of presenting them. He has chosen to give the reader only such information as Lee himself was able to obtain from day to day and hour to hour for this is the only way by which the reader can see the situation as Lee saw it. It has been no easy task, for it has required great care in disentangling the probable truth from conflicting testimony; but Freeman has done it with such kill that few will question his conclusions. He discards the story that Lee was able, by studying the personalities of his opponents, to predict what each one would do. On the contrary, Lee always insisted that one must expect the enemy "to do what he ought to do." Lee's method was to seek out every bit of information he could procure, weigh it, balance one thing against another, discard what was improbable, and then decide what was best to do with the means available. He saw his problem as a whole and was never confused by details. It is really exhilarating to watch, through the medium of these pages, the precise working of Lee's mind even in "the fog of war." When he made errors he discovered that they were errors and avoided repeating them. He devised new methods of meeting new conditions, as in his development of field fortifications not merely for the greater protection of his thinning ranks but also to hold a position with fewer men in order to gain freedom for maneuver with the others. Always he was painfully hampered in transportation facilities, in the commissariat, in the scarcity of clothing and shoes for his men, by the longer range and heavier metal of the Federal artillery, by the supreme difficulty, after the death of Jackson, of finding higher officers with the tactical skill to carry out his plans and at the same time to make wise use of the discretion he wished to give them. Step by step through the campaigns and reorganization and ever-increasing difficulties that Lee faced the author takes his readers. At the end of each major campaign he submits a clear, candid, critical review of Lee's operations.

On many difficult or disputed questions he throws new light, but only a few instances can be mentioned here. He justifies Lee for going into Maryland after Second Manassas because he could not feed his army where it was and the alternative was to retire behind the Rappahannock and leave an important section to the enemy. The decision to fight at Sharpsburg came only after he knew Jackson was at hand and he found the ground favorable for defense. One of Lee's greatest difficulties in the Pennsylvania campaign was the fact that two of the three corps of his army were under new and untried commanders, Ewell and A. P. Hill. His failure to get all his forces in front of Grant at the beginning of the Wilderness fight was because he had had to guard against a thrust down the railroad on his left. He was fully aware of the possibility that Grant might cross the James and strike at Petersburg before that movement was begun, but he could get no definite information, even from Beauregard, as to what corps of Grant's army had actually crossed until it was almost too late.

In a notable chapter in the last volume, Freeman sums up Lee's qualities as a commander in these words: "The accurate reasoning of a trained and precise mind is the prime explanation of all these achievements. Lee was pre-eminently a strategist, and a strategist because he was a sound military logician. . . . These five qualities, then, gave eminence to his strategy—his interpretation of military intelligence, his wise devotion to the offensive, his careful choice of position, the exactness of his logistics, and his well-considered daring. Midway between strategy and tactics stood four other qualities of generalship that no student of war can disdain. The first was his sharpened sense of the power of resistance and of attack of a given body of men; the second was his ability to effect adequate concentration at the point of attack. even when his force was inferior; the third was his careful choice of commanders and of troops for specific duties; the fourth was his employment of field fortification."

Among the mistakes of Lee, Freeman cites his too elaborate strategy in the Seven Days, his overestimate of the endurance of his infantry and his underestimate of the time required for the reduction of Harper's Ferry in the

Maryland campaign, his permitting Longstreet to stay so long in Suffolk in April. 1863, his selection of Ewell to command the Second Corps after Jackson's death, his acquiescence in the occupation of the Bloody Angle at Spotsylvania and the withdrawal of the artillery from that point, his "excessive amiability" at times when he should have been stern. But these errors weigh lightly against his supremely positive qualities.

Lee's relations with Jefferson Davis and his hold upon his men and the southern people are not hard to understand. He had no real difficulty with the Confederate president, partly because he understood him and had acquired a mental ascendancy over him and partly because he had a genuine respect for the civil authority and for Davis personally and was always tactful and deferential. Davis, moreover, had implicit confidence in Lee and always sustained him. His men knew that he looked after their welfare with assiduous care and that they could approach him without fear. Stories of his personal kindness to humble privates spread through the army and aroused affectionate reverence, while his successes against heavy odds developed the belief that he was invincible. To the people in general his successes and his character made him seem a leader raised up for them by divine favor.

Freeman refuses to make comparison between Lee and other great commanders of history on the ground that differences of conditions were so incommensurable that comparisons would be futile. One cannot but wish, however, that he had discussed the statement of certain recent military writers that Lee never showed that he was fitted for supreme command over a wide area such as

Grant exercised after March, 1864. While it may well be answered that Lee was never given such authority until it was too late to effect anything, a careful study of his correspondence between March 13 and June 1. 1862, while he was Davis' adviser—though with little real authority—should throw some light upon this question.

Dr. Freeman's delightful account of Lee's five years in the presidency of Washington College reveals the general as an educational leader. Not only did the trustees under the stimulus of his zeal rehabilitate the school materially and financially, but the faculty, under his guidance and in keeping with his anxiety for the training of southern youth in practical affairs, greatly enlarged the curriculum, anticipating many of the developments of later days. Meanwhile Lee, although greatly disturbed by the radical policy of reconstruction, kept studiously aloof from political or sectional controversies while doing all in his power to bring about eventual reconciliation between North and South. His prestige in his own section was as great as ever and no doubt much of the growth of the college was incident to his immense popularity. But his health had failed rapidly. A throat infection in March, 1863, followed by pericarditis had developed into what was probably angina pectoris. He died on October 12, 1870, in the midst of plans for the further development of the college.

In a final chapter, "The Pattern of a Life," Freeman tells simply but eloquently the manner of man that Lee was—his daily routine, his method of work, his simple and sincere religion, his kindliness and his humility. "Those who look at him through the glamour of his victories or seek deep meanings in his silence will labor in vain to make him appear complicated. His language, his acts, and his personal life were simple for the unescapable reason that he was a simple gentleman."

The four volumes contain numerous photographs and sketch maps. The reader who is not familiar with the geography of Virginia and other areas in which Lee operated will sometimes wish for a larger map. As the first two volumes came from the press several months before the last two, each pair is provided with a separate index—in the second and fourth volumes. There is also a "short title" bibliography, for which there seems little need, in the same volumes and a longer, most excellent critical bibliography filling twenty-seven pages at the end of volume IV. The mechanical work is faultless, the binding is handsome, and the work as a whole is worthy of its subject.

Charles W. Ramsdell

University of Texas

The Civil War and Reconstruction. By J. G. Randall. (Boston: D. C. Heath and Company, 1937. Pp. xvii, 959. Bibliography, illustrations, maps, graphs. $5.00.)[356]

Professor Randall has written a precise, lucid, and thoughtful book, packed with information and conclusions derived both from his own careful researches and from the work of other students. Although he has given relatively little space to the Reconstruction period—hardly more than one fifth of the text—he has supplied the long-felt need for a single scholarly volume covering both the war and the resulting radical domination of the conquered South.

The first three chapters, one of which deals with the institution of slavery, sum up the characteristics of Southern and Northern life; the next three trace the rise of sectional antagonism through the secession of the cotton states and the formation of the Confederate government. In the seventh, entitled "Buchanan's Dilemma," Randall shows more understanding of the peaceful aims and cautious policies of that harassed President than have most historians. Then follow two interesting chapters on the development of Lincoln's policy, as it culminated in the Fort Sumter affair, and the

[356] This is the review exactly as it appeared in *The Journal of Southern History,* Vol. 4, No. 4 (Nov., 1938), 532-533.

frightful plight in which it placed the Upper South. Randall condemns the shortsighted statesmanship which failed to prevent a war which he thinks was unnecessary. But while he seems to hold Lincoln guiltless of provoking hostilities, he asserts that the "April policy" was "unfortunate" in that it made war inevitable and also drove four more states over to the Confederacy. These two conclusions seem inconsistent, unless he means that Lincoln innocently blundered.

It takes 258 pages to bring us to the opening of hostilities. The next 430 pages are devoted to the progress of the war itself—with the emphasis upon the political, constitutional, and administrative problems—and constitute the heart of the book. Here Randall is at his best, especially in his account of developments in the North, for it is a field in which he has already done distinguished work. Too much praise cannot be given to the lucidity with which he analyzes complex constitutional problems and political situations and the admirably objective manner in which he presents them. If he is severe toward the radical opposition to Lincoln, most of us will agree that he has reason so to be. Less familiar at first hand with the situation in the Confederacy, he has been obliged to depend largely upon the writings of others in that field. If he misses some developments, it is rather the fault of those who have not published their findings than his own.

To the twelve years of Reconstruction only 191 pages are allotted. Possibly for the sake of unity, the author has restricted himself to the story of the contest between Andrew Johnson and the Radicals, a brief account of "The Grant Era," and the collapse of the radical governments in the South. The rising economic-social problems which became so important in later years he leaves untouched. On the constitutional aspects of the radical legislation he has made characteristically pertinent observations. His sympathies are clearly with Johnson and the victimized South, but he adds nothing to what other writers have said about Southern conditions.

The book is remarkably free from errors, but a few have been noted which should be corrected in later editions. Bluffton (p. 117) was not a man but a place in South Carolina; Thomas Bragg was not from Alabama (p. 342) but from North Carolina; under the Confederate "produce loan" of 1861 the treasury did not receive the produce itself (p. 351) but only the proceeds from its sale by the planter who was pledged to invest the money in Confederate bonds; the Southern state supreme court judges displaced by military officers under the Reconstruction Acts (p. 754) had not been appointed "under the Confederacy." Led astray by John Bigelow's *Retrospections,* the author has Governor Ellis of North Carolina say that his state would secede because of Lincoln's election (p. 184, n. 1), when in fact Ellis expressed just the contrary opinion. (A correct copy of the original letter is in Nicolay and Hay, *Abraham Lincoln, A History,* II, 307-308.)

If the "general reader" has the weak, flabby mentality which reviewers sometimes attribute to him, he may find this book too heavy for his taste, despite its lucid style; but the intelligent reader who wishes to know what up-to-date scholarship has discovered about the Civil War and its disgraceful aftermath will find it fascinating. It is an excellent book to put into the hands of an advanced college class and doubtless it will be used widely in that way. Contributing greatly to its interest are the many illustrations: portraits, photographs, original drawings, posters, handbills, and documents. The bibliographical note and the extensive bibliography, totaling forty-three pages, are most helpful, except that the arrangement sometimes makes difficult the location of a particular item. The index is very satisfactory.

Charles W. Ramsdell
University of Texas

The Repressible Conflict, 1830-1861. By Avery Craven. (University, Louisiana: Louisiana State University Press, 1939. Pp. xii, 97. $1.50.)[357]

This delightful little book consists of the second series of the Walter Lynwood Fleming Lectures in Southern History which Professor Craven delivered at Louisiana State University in February, 1938. His general theme is that the Civil War was not inevitable but for the foolishness of men who let their emotions run away with their reason. If this conclusion is not entirely novel, the argument is presented in a manner both refreshing and stimulating.

Although the South, like other great sections, contained elements of unity, its interests and social attitudes were so diverse that its people never thought seriously of separate nationality until fused into something like unity by pressure from without. This pressure began with the abolitionist crusade which forced the Southerners into a defense not merely of slavery but, finally, of the whole economic and social structure of their section. The abolitionists never bothered to find out what slavery in the South was actually like: they began with the assumption that it was wholly evil and the slaveholders depraved and launched into a perfect ecstasy of vilification. The incensed Southerners replied in kind. As

[357] This is the review exactly as it appeared in *The Journal of Southern History,* Vol. 5, No. 4 (Nov., 1939), 553-554.

the campaign of exaggerated accusation progressed, sectional economic and political rivalries became linked with the contest over slavery and added more clamor to the din of mutual recrimination. Thus the leaders of the industrial groups of the Northeast, who had been contending with the agriculturists of the South and West since the days of Hamilton and Jefferson, saw in the slavery issue an opportunity to split the agrarians, whose ablest leaders were Southern planters, by arousing the West against the South. They therefore aided the "holy war" against the Southern "aristocracy," less because they disliked slavery than because they wished to break the political power of their opponents. By the time the North was sufficiently united on this program to elect Lincoln in 1860, the campaign of hate had gone so far that reconciliation was impossible. "God's purposes" must be fulfilled by a bloody war.

The three lectures, "Foundations of Southern Nationalism," "The Peculiar Institution," and "The Repressible Conflict," are well articulated and, although there is necessarily some repetition, they flow along in logical sequence. While the chief interest of the special student will be in the interpretation, the general reader who has derived his knowledge of the period from textbooks and the older "standard" histories will find the evidence a revelation and will relish the delightfully easy and informal style. Teachers who have struggled with the intricacies of the subject will be grateful for Craven's gift for luminous exposition, for his ability to reveal in simple terms the essence of things which many have sensed but found hard to make clear to their students. For instance, his analysis of "the peculiar institution" is an excellent example of how informed common sense may be applied to historical interpretation. The reader feels that the situation described was a perfectly natural one for its time. Such art is possible only when a discerning mind is sustained by ripe scholarship.

The little book is beautifully printed and is a credit to the Louisiana State University Press. Because it is published primarily to make the lectures available to those who were not privileged to hear them, the usual footnotes and index have been omitted.

Charles W. Ramsdell
The University of Texas

The American Civil War: An Interpretation. By Carl Russell Fish, edited by William Ernest Smith. (New York: Longmans, Green and Company, 1937. Pp. xi, 531. Bibliography, illustrations. $3.50.)[358]

For many years before his untimely death in 1932, Professor Fish had been planning two volumes, one on the Civil War and the other on Reconstruction, which he hoped to make the crowning work of his life. Death intervened when he had completed the fourteenth chapter, which ends with the first annual message of Andrew Johnson, but before he had had time to finish the work of revision. Professor Smith, one of his former students, has prepared these chapters for publication, written two additional ones on governmental .finances and the constitutional aspects of the period, and has added an extensive bibliography.

What Fish undertook to do was, first, to disentangle, describe, and evaluate the multiple factors which divided the country under two governments and brought on war, and then to do the same thing for those factors which determined the course and outcome of the conflict. He

---

[358] This is the review exactly as it appeared in *The Journal of Southern History,* Vol. 4, No. 4 (Nov., 1938), 533-534.

has neglected no important factor and his attitude is as impartial as the most conscientious scholar could wish. More than this, he reveals an understanding sympathy for the convictions of honest men and women on both sides of the battle line. His central figure is, inevitably, Lincoln; and he is perhaps at his best in analyzing Lincoln's apparent shifts of policy—which he regards as mere maneuvers—while he adhered steadfastly to his central purpose of restoring the Union with as little damage as possible to any portion. The book abounds in eloquent, quotable passages, in vivid pen pictures of individuals, in penetrating observations, in flashes of humor.

It is interesting to compare this book with that of Randall. In general point of view they are much alike, but in methods of attack they are sharply different. Randall's has a broader scope, even for the shorter period covered by Fish, and though much more detailed is generally more compact. Fish's work, though not lacking in detail, is more of a brilliant and informed running commentary on the war and the men who waged it. In fact, the two books supplement each other admirably. Fish paid but little attention to the difficult constitutional issues which worried lawyers and courts as well as many other individuals or to government finances. It was to supply this obvious lack that the editor has supplied a chapter on each of these subjects. While in style they fall far short of the brilliancy of Fish and contain some statements to which the reviewer takes exception, they are excellent summaries of difficult subjects and are packed with much useful information. Professor Smith explains in his preface that the original chapters were in a handwriting "not as easy to read as Horace Greeley's" and that, because they had not been revised by their author, it was necessary to rewrite many paragraphs and pages and to rearrange sentences. These things as well as the inevitable typographical error may account for some of the numerous slips that occur, as in the initials of proper names, wrong dates, an occasional cloudy sentence, and the substitution of a wrong word for the one obviously intended. An example of this last is the word "improvement" (p. 171) for "impressment." Twice Pollard is said to have been the editor of the Richmond *Enquirer* (pp. 127, 173) when, of course, his paper was the *Examiner*. Space does not permit a complete listing. Unfortunately, Professor Fish's original footnote citations were lost and the task of replacing them proved too much for the editor. While their absence is a serious loss, Dr. Smith has to some extent repaired the damage with a bibliography of forty-three pages. The index is useful, if not complete.

Charles W. Ramsdell
University of Texas

State Rights in the Confederacy. By Frank Lawrence Owsley. (Chicago: The University of Chicago Press, 1925. xi + 288 pp. $2.50.)[359]

Ever since Appomattox, students of the War for Southern Independence have sought to fix upon the chief causes of southern defeat. In earlier years the credit was about equally divided between Providence and the heavier battalions; more recently attention has shifted to the industrial and social differences between the warring sections. Some earlier southern writers—chiefly anti-Davis politicians—insisted that the despotic measures of the Confederate government, the invasion of the rights of the state governments and of the people, broke down public confidence and support. Professor Owsley takes the opposite position: namely, that state rights jealousy and particularism so weakened the general government that defeat, which otherwise would have been almost impossible, became inevitable. He sums up his thesis in this paraphrase of Jefferson Davis' well-known epigram: "If a monument is ever erected as a symbolical gravestone over the 'lost cause' it should have engraved upon it these words: 'Died of State Rights.'"

In support of this thesis, he endeavors to show that through their governors, legislatures, or courts the several

[359] This is the review exactly as it appeared in *The Mississippi Valley Historical Review*, Vol. 14, No. 1 (June, 1927), 107-110.

states, practically without exception, withheld from the Confederate government during the first year of the war both the arms and the men necessary to immediate success; that they sought to defeat the operation of the conscription acts; that they constantly interfered with the control of their troops in the Confederate service; that they fought against such necessary measures as the suspension of the writ of habeas corpus and the impressment of supplies for the army. Of course it is not at all difficult to find evidence that state authorities frequently hampered the Confederate government, and Professor Owsley has found it. To the extent that he has brought this material together, the book—which is well organized and very well written—is a useful contribution to an interesting and perplexing problem.

The difficulty is that he has tried to prove too much and has laid his book open to severe criticism in a very important particular, the handling of evidence. Possibly the author became over-enamored of his thesis and, like other lovers, lost something of his critical powers; or it may be that he was more interested in a vivid and readable narrative than in painful accuracy. At any rate, he has accepted isolated and casual statements as bases for sweeping declarations; he has read into some of his sources statements that are not there even by implication; and he has ignored evidence that tends to disprove or to qualify materially portions of his general thesis. These faults are most frequent in the first two chapters, but they are found throughout the book. Since only a few illustrations are possible here, let us consider the treatment of the two governors against whom it is easiest to make out a case for willful obstruction—Joseph E.

Brown of Georgia and Zebulon B. Vance of North Carolina.

Now, it is a matter of common knowledge that in his relations with the Richmond government, Governor Brown was often a cantankerous nuisance, though rather an able one; and this reviewer has no ambition to appear as his defender. But the historian must be fair even to Joe Brown. There seems to be no ground for the assertion (p. 15) that Brown, not content with holding on to the state's arms, was determined "to get as much more as possible out of the Confederate government"; nor, with reference to the matter of the state's powder in Augusta (p. 16), that he was willing to break up the whole system of defense worked out in Richmond. Nowhere is Brown credited with what he actually did to raise troops for the Confederate service and to supply them with arms and munitions. He is represented as asking Secretary Benjamin (p. 19) for arms "to equip four regiments for local defense," when in fact these four regiments were already in camps of instruction and mustered into Confederate service. The account (p. 21) of the seizure of arms from the steamer *Bermuda* by the Confederate General Lawton and of Brown's alleged complicity therein, is a strange reading of plain evidence. Brown is made responsible both for the rumors of invasion by way of Brunswick and for Lawton's seizure of the arms, although the very documents cited show that the Secretary of War himself had warned Lawton to look out for an attack there. What is there in Benjamin's letter of reproof to Lawton, or in anything else, to show that the Secretary of War "though the hands were those of Esau [Lawton], recognized the voice of Jacob [Brown]"? In the quotation from that letter the words "by its officers" are omitted after "appropriated" with the effect of changing the meaning.

Every charge of withholding arms from the general service made against eight other states (pp. 10-15) is based upon incomplete evidence, or rather *ex parte* evidence. The conclusion that the states in the aggregate withheld, in 1861, from 200,000 to 350,000 stand of arms (pp. 22-24, 272-73), is based upon a series of assumptions of more than doubtful value. Though Professor Owsley admits nothing of the sort, there is plenty of available evidence that many of the states armed a large proportion of the volunteers which they sent into the service at the beginning of the war. In fact, some of the governors went beyond their constitutional powers in extending aid to the Confederacy; for they had no legal authority to transfer or to sell state arms and equipment to the War Department when neither convention nor legislature had provided for it.

Like so many other writers in this field, Professor Owsley regards Governor Vance as a constant obstructionist, a consistent enemy of the administration at Richmond. It really does not require much knowledge of conditions in North Carolina and of Vance himself to see the injustice of this view. No one knew better than the young governor how rapidly disaffection was spreading in the western part of that state, and he wished to remedy the situation; but the Richmond authorities could not make the exceptions to their general policies which he thought necessary. Vance was high spirited and too impetuous, and at times he was clearly wrong; but there can be no doubt of his sincere desire to further the southern cause. He did not say to the legislature (p. 39) that "he had been tricked" by the Confederate government, nor anything like it; and no such words as "without interference" (p. 40) appear in his polite letter to Davis, November 25, 1862. His controversy with Davis in 1864 was both unnecessary and unprofitable, but to call his letters a "torrent of abuse" (p. 178) seems an extravagance. To brand his suggestion that he be allowed to enlist some 1,200 deserters as an "audacity" (p. 44) hardly seems justified in view of the fact that Assistant Secretary of War Campbell approved of it. Neither is it correct to say that Vance "swung the state into a position of violent opposition to the suspension of the writ of habeas corpus" (p. 173); nor that the people of North Carolina would have borne impressment for a long time had Vance supported the Confederate government (p. 248). Vance was too good a politician not to conform outwardly to a popular feeling as strong as that in North Carolina on these subjects. Yet there is manuscript evidence that Vance hoped to see Chief Justice Pearson's "in chambers" opinions in certain habeas corpus cases reviewed and reversed by the full court in order to prevent the breaking down of the conscription act. His protests against impressments came only after the conduct of agents and officers had become a scandalous oppression. Indeed, the widespread opposition to impressment was due to discontent with the manner of its enforcement rather than to abstract state rights theories.

Professor Owsley's condemnations are sweeping. His rogues' gallery is crowded with governors. In the midst of this dismal spectacle of universal gubernatorial wrong-headedness, it is cheering to behold the noble Governor

Milton of Florida administering a dignified rebuke to the irascible Brown (p. 114); but alas! a little later (pp. 204, 243) even the good Governor Milton is caught opposing the impressment agents and guiltily "sawing wood" about conscription. Lubbock of Texas is placed among those troublemakers and opponents of Davis who "rather than give up their theory preferred to see the whole Confederacy go down in defeat" (p. 228). Lubbock, it happens, regarded himself as a staunch supporter of the administration; and Davis also must have so looked upon him, because immediately after the expiration of Lubbock's term in 1863 the President took him to Richmond as a member of his staff and advisor on Trans-Mississippi affairs.

Never once throughout the book does the author suggest that any policy or act of the Confederate government or of any high Confederate official could be mistaken. Never once, except when Milton rebukes Brown, does any governor do a wise or generous thing. There is never a word of appreciation of their constructive work nor of the difficulties with which they had to contend. Nor is there anywhere any recognition of the fact that not a single state supreme court failed to sustain the constitutionality of the conscription acts and of the acts suspending the writ of habeas corpus. The errors are most conspicuous in the accounts of conditions during the first two years of the war; for, after Gettysburg and Vicksburg, increasing weakness and loss of confidence produced so much discontent, demoralization, and recrimination that the facts better fit the thesis.

The format of the little book is pleasing, but there are a few typographical errors. When Judge Pearson discharged the soldier (p. 175), he was sitting "in chambers," not in a place by that name. "Northup" (p. 242) should be "Northrop." Some footnote references and some titles are wrong, but such errors are hard to eliminate by the most careful proofreading. There is no general bibliography. The Index is scanty; it is inadequate both for proper names and subjects.

Charles W. Ramsdell

Life and Labor in the Old South. By Ulrich Bonnell Phillips. (Boston: Little, Brown and Company, 1929. xix+375 pp. Illustrations and maps. $4.)[360]

The appearance of Professor Phillips' latest book was greeted with eagerness by all students of southern history, not because it had won a large cash prize as "the best unpublished work on American History," but because so much was expected of one who enjoys unquestioned preeminence in his field. The high expectations with which the book was awaited prove, on the whole, amply justified; but many other specialists in southern history will feel disappointment over certain omissions, some of which will be indicated later in this review. Specialists, of course, are notoriously hard to please.

Professor Phillips has here attempted what no one has ventured before. He has sought to present a continuous picture of the stream of life which for more than two centuries poured through that diverse group of sections known collectively as "the South." It is a large theme for a relatively small book; and the extent of the project raises at once the question of how he has handled the problems of proportion, condensation, inclusion, and exclusion. Before considering those features it should be said that every page bears evidence not only of the

[360] This is the review exactly as it appeared in *The Mississippi Valley Historical Review*, Vol. 17, No. 1 (June, 1930), 160-163.

author's intimate and accurate acquaintance with his subject but also of his objective though sympathetic attitude in presenting the ways of life of the southern people. He has blended skillfully the chronological order with the topical, but tends rather to follow the latter, for he turns back repeatedly to illustrate a point by reference to some colonial instance. He has condensed his account, where necessary, with an expert hand, but he has enforced every important statement by a wealth of apt quotation from individuals who are made to speak for others as well as for themselves. This device, along with the author's easy style and never-failing humor, imparts a lighter note and a more piquant flavor than is usually found in solid historical work. As a result the book is never heavy reading, and many persons who ordinarily avoid learned historical volumes have testified to the pleasure they have derived from this one.

The first five chapters, about one-fourth of the volume, are devoted to an excellent description of the southern topography, climate, and soil, and to charming summaries of the effect of environment upon the settlement of Virginia, Maryland, the Carolinas, and Georgia, and of the advance from the colonial back-country across the mountains into the bluegrass of Kentucky and Tennessee and on into Missouri. The next ten chapters, the heart of the book, comprising more than two hundred and thirty pages, describes the development of the plantation regions, their mode of life, and their problems, the handling of the great staple crops, the economic and social aspects of slavery, and typical planters of Virginia, the Southeast and the Southwest, and their overseers. In these chapters, as was to be expected, Phillips has drawn heavily upon material used in his earlier books, especially his *American Negro Slavery,* but he has also produced new evidence. One chapter is then given to a description of southern homes of all classes and conditions. The last two chapters describe the characteristics, respectively, of "the plain people" and "the gentry." In addition to more than forty photographs, chiefly of southern houses, there are three plantation maps, a graph of slave prices, and a general economic map which shows the distribution of staple crops and the lines of transportation by river and railway in 1860. It will be seen that Phillips has limited himself almost wholly to what has long been his major interest, southern agriculture and agricultural society—especially the plantations and plantation economy, the masters, and the negro slaves. Since the next volume promised is to be restricted to "the course of public policy," one might reasonably have expected in the present one a more comprehensive treatment of all the important developments in the economic as well as the social life of the old South. But there is practically nothing about such industries as mining, lumbering, manufacturing, and milling, the decline of the import trade, or the ambitious schemes behind the building of railways. The spread of railways in the eastern cotton belt he has described in an earlier work, but surely the subject deserves more space than the two pages (146-48), however excellent the summary, which he has allotted to it here. Planters, farmers, overseers, or slaves appear on nearly every page; but the manufacturers, iron-masters, merchants, railway builders, factors, bankers, and educational leaders, who were familiar to their planter contemporaries, are omitted. It is hardly a measure of their importance to give to the six million non-slave holders, comprising three-fourths of the white population, only one short chapter of fifteen pages.

Probably every investigator in the field of southern history has several questions to which he hoped Professor Phillips would suggest an answer. Was slavery over a long period economically profitable? Were the small farmers really oppressed by the institution of slavery, especially those outside the denser plantation regions? Barring the cataclysm of the war, what does the evidence show as to the probable future of the institution ?

What had been accomplished by the movement for agricultural reform in the eastern cotton belt by the end of the eighteen-fifties? To what extent, if at all, had southern agriculture been affected before 1860 by the increasing financial power of the East? It is regrettable that he nowhere gives explicit answer to any of these queries, for if anyone could answer with authority, it should be Phillips. A reviewer must not be too persistent with his questions, but one more may be ventured. Is it not time we were arriving at some fairly workable synthesis of the South as a whole just before the catastrophe of the sixties? Perhaps it doesn't matter; but there are some who hope that in the next volume, even though it is to be concerned primarily with politics, there will be room for a broader synthesis.

Notwithstanding the qualifications here suggested, there can be no question of the value of this delightful and illuminating book. It will not only be a boon to teachers and their students in college courses in southern history, but it will give lasting pleasure as well as a better understanding of the old South to a wide circle of nonprofessional readers.

Charles W. Ramsdell

The Story of the Confederacy. By Robert Selph Henry. (Indianapolis: The Bobbs-Merrill Company, 1931. 514 pp. Illustrations, maps, bibliography, and synoptic table of events. $5.00.)[361]

This book is intended for the general reader rather than the special student. The author modestly admits (p. 494) that "of the research of the historian, that is, digging out the hitherto unknown, there is little in the book"; but it must be said for him that he has made good use of a rather wide range of secondary materials as well as of the *Official Records of the Union and Confederate Armies.* Although he discards footnote references, the reader who is familiar with the literature of the subject will be able to identify most of the sources used. Since he is concerned primarily with the military operations, Mr. Henry touches but lightly on political controversies, economic and social conditions, public finances, or constitutional problems. As a military narrative, the book is clear, concise, well-balanced, and judicial. It is, in fact, one of the best accounts yet written in one volume of the whole scope of military operations from Pennsylvania to Texas. The author seem to disclaim "impartiality of spirit," but it will be hard to find evidence of prejudice in his story. His

---

[361] This is the review exactly as it appeared in *The Mississippi Valley Historical Review*, Vol. 18, No. 4 (March, 1932), 587-588.

experience as a railroad official (he lives in Nashville) has caused him to give more attention than most writers have done to the important part played by the railroads in the movement and supply of the armies. Some of his judgments are interesting, as, for instance, that September 17, 1862—the day when Lee was checked at Sharpsburg and when Bragg, at Munfordville, Kentucky, decided to go to Frankfort instead of to Louisville—marked the real beginning of the decline of the Confederacy. Sketch maps enable the reader to follow the general movements of each campaign. Detailed battle maps are not furnished, but descriptions of the local topography are so clear that they are not really necessary. The illustrations consist of portraits of Confederate generals and photographs of southern capitols and of Richmond and Charleston. The bibliography, strangely enough, is almost wholly confined to books by or about Confederate leaders.

Charles W. Ramsdell

Constitutional Problems under Lincoln. By James G. Randall. (New York and London: D. Appleton & Co., 1926, pp. xviii, 680.)[362]

Notwithstanding the interest which students of American constitutional history have long taken in the difficult legal problems which confronted the United States Government during the Civil War, Professor Randall is the first to undertake a careful and comprehensive study of the whole subject. It is a task full of difficulties, but the reader will not get far into the book without becoming convinced that the author is admirably equipped for the work. Gifted with judicial temper and clearness of perception, he has acquired a thorough acquaintance with the background of his problems and a mastery of the technique of the law involved in them. The book is the fruit of fifteen years of research which extended through not only the printed sources but also into the manuscript records of courts and departments.

The administration of Lincoln faced constitutional problems such as had never before risen to perplex a government. It was difficult to find precedents for a crisis of such dimensions. The central problem was to maintain executive efficiency, in order to save the Union itself, without injuring permanently constitutional government.

---

362 This is the review exactly as it appeared in *The Southwestern Political and Social Science Quarterly*, Vol. 9, No. 3 (December, 1928), 357-359.

That this was done in the end, despite some lapses and questionable expedients, was due, so Randall thinks, to the fundamental legal-mindedness of the people and to the caution and moderation of Lincoln himself.

It is impossible in a brief review to indicate adequately all his major conclusions. The introductory chapter, which reveals the mind of the historian, shows how social forces condition constitutional growth and existing conditions affect judicial decisions. Secession was, at bottom, a political rather than a constitutional question. The "war powers" in relation to the Constitution and the legal nature of the war itself are then examined. Here was one of the major difficulties, for the Government was forced to adopt a dual theory: first, that the Confederates were rebels and traitors; second, that they were belligerents with belligerent rights. The first was followed in argument and declamation, but the second had to be followed in general practice. This is most clearly shown in the case of captives, both on sea and land, and in the treatment of paroled prisoners after the surrender of the Confederate armies. The attempt to apply the law of treason to Jefferson Davis was almost the sole application of the first theory under the criminal statutes, and it was abandoned from fear of failure. Three chapters deal with the suspension of the privilege of the writ of habeas corpus, military rule, arbitrary arrests, and military commissions. As to whether the President or Congress has the constitutional power to suspend the writ of habeas corpus, Randall regards the Civil War precedents as inconclusive; for though he inclines to the older view that the power belongs to Congress, he points out that under stress of a similar crisis the example of

Lincoln may be followed again. As to the use of martial law and arbitrary arrests by mere presidential order, he admits that Lincoln assumed powers that amounted to a dictatorship, and that the acts of Congress were powerless to restrain him; but he shows that after all it was a mild dictatorship without intent to subvert the Constitution. The Indemnity Act of 1863, and the amendment to it in 1866, which was intended to protect officers from prosecution for violation of private rights, he thinks a very unsatisfactory measure though the Supreme Court sustained it on the score of constitutionality. Conscription by the general government he regards as clearly constitutional. Three chapters are given to the subject of confiscation. He points out that these acts involved an extraordinary extension of the doctrine of belligerent powers, questions the majority opinion of the court which upheld them, and shows how easily corruption entered into their administration. Lincoln's insistence upon the constitutional restraints forced a construction of the act which resulted in the restoration of most of the confiscated property. The chapter on emancipation emphasizes Lincoln's cautious and conservative approach to the problem and contains some interesting information concerning J. Q. Adams's arguments against emancipation under the war power after the British had resorted to it in 1814. One chapter describes military rule in the occupied districts of the South; one is given to state and federal relations in the North; one to the partition of Virginia, for which he finds no just reason; and another to the attitude of the Government toward the press. The last chapter sums up his findings. In the light of English and American constitutional principles Lincoln's government was conspicuous for its irregular and extra-legal methods; the dual nature of the war caused inconsistency in legislation and administration; the Supreme Court exerted no effective restraint upon the President or upon Congress; but there was no permanent damage to civil liberty. He draws an interesting contrast between the methods of Lincoln and Wilson during war-time, showing that the latter, unlike Lincoln, was careful to have the authorization of Congress for the exercise of extraordinary powers.

Because of its spirit of detachment, its penetrating analysis of the tangled legal issues involved, and its lucidity the book is well adapted to college classes. It supersedes all other works on the subject. Its usefulness is further enhanced by an excellent bibliography and a full index.

Charles W. Ramsdell
University of Texas

Guide to the Study and Reading of American History. By Edward Channing, Albert Bushnell Hart, and Frederick J. Turner, professors in Harvard University. (Boston and London: Ginn & Co., 1912. Pp., xvi, 650.)[363]

The present volume is a great improvement over the first edition, which was published in 1896. It has been brought down to date; it gives references to more available books; and it enlarges the sections on social, economic, and industrial history, making them especially valuable. Professor Turner, who was not connected with the earlier edition, has contributed many valuable references to writings on Western history. The whole work, however, has been done over, and will be found very helpful to students and teachers in every field of American history.

Charles W. Ramsdell

---

[363] This is the review exactly as it appeared in *The Southwestern Historical Quarterly*, Vol. 16, No. 3 (January, 1913), 334.

Bedford Forrest, The Confederacy's Greatest Cavalryman. By Captain Eric William Sheppard. (New York: The Dial Press, 1930. 320 pp. Plates and maps. $5.00.)[364]

The interest of British military men in the careers of American soldiers of the Civil War seems to increase rather than to diminish. Captain Sheppard, of the Royal Tank Corps, now adds his name to the list of Forrest biographers—Thomas Jordan and J. P. Pryor in 1868, Dr. John Allan Wyeth in 1899, and J. H. Mathes in 1902, all ex-Confederate soldiers. The last two produced careful and scholarly works; but since no Englishman heretofore has written about the great cavalryman, Captain Sheppard regards his new book as turning "practically virgin soil." But he has found little or nothing new of any importance—for Wyeth, especially, left little to glean—nor has he given a new interpretation of Forrest's military career. Indeed, he expressly disclaims any desire to describe his hero's work "as a covert means for instruction in the military art." He has succeeded in portraying both the man and the soldier.

A bare recital of the exploits of this untutored and picturesque genius, son of a cross-roads blacksmith, slave-dealer and planter, whose demonic energy, reckless

---

[364] This is the review exactly as it appeared in *The Mississippi Valley Historical Review*, Vol. 18, No. 1 (June, 1931), 95-96.

courage, and instinctive perception of just what should be done in any emergency raised him finally to a lieutenant-generalship, could not be read without amazement. In Captain Sheppard's vivid narrative Forrest becomes, as he doubtless appeared to both his own men and his enemies, something almost superhuman. He not only managed to win his fights with inferior numbers and equipment, but he overcame countless obstacles of rainy or wintry weather, unbridged streams, all but impassable roads, and failing supplies, to strike where he was least expected. Always careful of the condition of men and horses, he knew how when the need arose to drive them to the very limit of physical endurance. Ignorant of formal tactics, his tactical arrangements for battle were models. Careless of minor matters of military form and decorum, he was the sternest of disciplinarians in all that was essential. Fond of the clash of personal combat and of risking his own life at the head of a charge, he never lost sight of his objective or threw away in excitement the direction of his battle. Serving until near the end of the war as a subordinate of men who looked upon him as only an audacious and successful raider, he saw strategic possibilities more clearly than most of his official superiors, but was never given a chance to put his larger conceptions into execution until his resources were gone. In order to appeal to the general reader, Captain Sheppard has introduced many imaginary conversations and two or three fictitious characters. He has thereby undoubtedly added color to his narrative; but, although in his preface he has honestly confessed these embroideries, those readers who prefer fact to fiction will be irritated by them. Despite all this, it is evident that he has examined the sources with care—although he has eschewed all footnote citations—that he is thoroughly familiar with the western campaigns, and that in all important matters he has followed the records. Here and there are a few minor errors, and an occasional typographical slip. The style is rapid, vivid, and clear, but is marred now and then by such unusual and unnecessary words as "spate" and "opted"—the latter in wearisome reiteration. Two large maps and four battle plans enable the reader to follow the movements of both Confederate and Union forces. At the end is a good brief critical bibliography and a satisfactory index.

Charles W. Ramsdell

SOUTHERN EDITORIALS ON SECESSION. Compiled by Dwight Lowell Dumond. (New York: The Century Company. 1931. Pp. xxxiii, 529. $4.00.)[365]

Southern Editorials on Secession, the first volume to be published under the direction of the American Historical Association from the income of the Albert J. Beveridge Memorial Fund, contains 183 editorials which ran from January 6, 1860, to May 9, 1861. All but four, which are from one paper in St. Louis, are taken from newspapers east of the Mississippi. Delaware, Maryland, Florida, Arkansas, and Texas are not represented. The reason for these geographical gaps doubtless is that the editor sought representative opinions of groups rather than of localities. Nevertheless, the geographical distribution of the selections is suggestive of the sectional weight of discussion: 99 editorials come from the cotton states; 56 from Virginia, North Carolina, and Tennessee; while 28 are taken from Kentucky and Missouri. The New Orleans papers are most fully represented, since 66 editorials, more than one-third of the total, are taken from the five journals of the Crescent City.

These editorials disclose a variety of opinion that

[365] This is the review exactly as it appeared in *The North Carolina Historical Review,* Vol. 10, No. 4 (October, 1933), 330-331.

must astonish those readers who come upon them for the first time. They range from those of such an uncompromising advocate of secession as the Charleston *Mercury* to those of unconditional Unionists like the Louisville *Courier*. Incidentally, they prove that the *Mercury* was far from being representative of general southern opinion. They show that the South throughout most of 1860 was divided into groups which disagreed upon almost everything—even upon the practical value of the protection of slave property in the territories. In fact, distrust and fear of the designs of the "Black Republicans" was the only thing common to them all. But not even after the free states had elected the anti-southern candidate could the southern men agree upon a plan of action. Soon, however, the march of events forced one group after another into line with the state rights extremists. The secession of South Carolina, the failure of the compromise measures in Congress, and the hostile and threatening attitude of the Republican press and Republican leaders turned more and more conservatives to secession. In the border states the pre-inauguration speeches of Lincoln, especially his speech at Indianapolis, made a bad impression; but it is interesting to note that some of the editors who in the end supported secession professed to see nothing alarming in his inaugural address. It was, of course, the call for troops after the attack on Fort Sumter that forced a choice between the hated Black Republicans and the lower South. It is instructive to trace the changes in attitude of such conservative journals as *The True Issue* of New Orleans, the Wilmington (N. C.) *Journal,* and the Nashville *Banner*.

One wishes that Professor Dumond had found space for a few representative newspapers of western Texas and Arkansas, where the frontier situation presented a peculiar problem; but for the editor's purpose the selections are admirably chosen. The book will be of great value not only to students, but also to those general readers who have maintained an interest in one of the most fateful periods of American history.

Charles W. Ramsdell
Austin, Texas

THE SECESSION MOVEMENT, 1860-1861. By Dwight Lowell Dumond. (New York: The Macmillan Company. 1931. Pp. vi, 294. $2.50.)[366]

In *The Secession Movement, 1860-1861,* Professor Dumond has not only capitalized the material gathered for the *Editorials,* but has also worked through a far wider mass of sources—congressional debates and reports, contemporary convention records, speeches, pamphlets, correspondence, and a considerable body of secondary accounts.

The first chapter is a very clear analysis of the several party positions on the leading sectional controversy, the protection of slave property in the territories, and contains the best explanation known to this reviewer of the reasons why the southern Democrats could not accept the views of Douglas on that subject. The author next relates the failure of the conservative elements to accept the proposal of South Carolina and Mississippi for a general southern conference to formulate a program or statement of principles, a failure that was to strengthen their opponents, the separate state actionists. Three chapters are given to the conflicts between the Douglas and anti-Douglas Democrats through the Charleston, Baltimore, and Richmond conventions. The detailed story

---

[366] This is the review exactly as it appeared in *The North Carolina Historical Review*, Vol. 10, No. 4 (October, 1933), 331-333.

of the ruthless parliamentary tactics of the Douglas delegates in applying the unit rule and in finally counting the bolting delegates for Douglas in order to give him a fictitious two-thirds vote is not only a distinct contribution to political history, but does much to explain the growing belief of southern Democrats that the gulf between the sections was too wide to be bridged. Although the Constitutional Unionists as well as the southern Douglas men are generally represented as anti-secessionists because they denounced the Breckinridge men as secessionists, Dumond shows that they were never unconditional Unionists, but were as determined as the Breckinridge followers to resist any aggression on the part of the Republicans. One of the most interesting chapters in the book explains the group attitudes toward secession. One group was the "immediate separate state actionists" who believed that the issue was already made up and who now opposed a southern convention because it meant delay, and they thought delay was dangerous. The more numerous cooperationists were themselves divided into three groups: (1) those who wanted immediate secession, but by united state action;·(2) those who wished to try first for compromise, and if that failed, to secede; and (3) those who preferred to wait for an "overt act" of hostility on the part of the Lincoln administration, and if that came, to secede. The last two were strongest in the border states, the others strongest in the lower South. The secession of South Carolina greatly strengthened the separate state actionists in the other cotton states and the refusal of the Republicans in Congress to agree to any essential compromise was decisive. The failure of compromise is laid squarely upon the Republican leaders. Major Anderson's removal to Fort Sumter and Buchanan's effort to reinforce the forts led to the seizure of other forts and arsenals by the state authorities and accelerated the secession of the Gulf states. The final chapters describe the futile efforts of the border state leaders to prevent collision and to bring about a "reconstruction" and compromise through the Peace Conference at Washington. The book ends somewhat abruptly with Lincoln's inauguration. Another chapter on the change of sentiment in and the secession of the border states would have better rounded out the work.

This book is by far the best yet written on this engrossing subject. A captious critic might object that the author has told the story solely "from the southern point of view," but it was necessarily that point of view which Professor Dumond had to explain. He relates what southern men thought as each stage of the crisis arrived and what they did; he nowhere assumes to pass judgment upon them. The thoughtful reader is apt to reflect that, given human nature as it is and the situation as it was, the result could hardly have been otherwise.

Charles W. Ramsdell
Austin, Texas

Aeronautics in the Union and Confederate Armies, With a Survey of Military Aeronautics Prior to 1861. Volume I. By Haydon F. Stansbury. (Baltimore: The Johns Hopkins Press, 1941, pp. xxii, 421. Illustrations.)[367]

In this work Dr. Haydon presents the first adequate account of the use of military balloons in our Civil War, a subject which has been strangely neglected in view of the prime importance of military aeronautics today. This first volume (there is to be at least one more) extends only to March, 1862, when the Peninsula campaign was beginning, and it deals only with the balloonists of the Union army.

European armies, especially the French, had been experimenting with observation balloons since 1794, but American officers had no experience with them. Nevertheless, a few civilian enthusiasts had been active and had made considerable progress both in the construction of balloons and in knowledge of air currents. When the call for troops came, several well-known balloonists offered their services to the War Department and were accepted. Unfortunately, however, the balloon service was never incorporated into the army

[367] This is the review exactly as it appeared in *The Southwestern Social Science Quarterly,* Vol. 22, No. 2 (September, 1941), 187-188.

establishment, for though balloons and equipment were purchased for the army the balloonists themselves were merely civilians hired at so much per day, without official status and subject to the orders of army officers some of whom were entirely ignorant of the peculiar needs of this service and more interested in established routine than aiding the balloonists. Some of the higher officers were scornful of the balloons and refused to use them; but others, such as McClellan, Porter and Butler gave intelligent support to the new service, and got very satisfactory results.

For observation of the enemy's positions and movements the aeronaut, sometimes accompanied by an officer and a telegrapher, went up in a captive balloon, moored as close to the front lines as safety permitted, to a height of from 1,000 to 5,000 feet. Telegraphic communication with headquarters was maintained by a wire from the basket along the cables to the mooring position. On clear days the observer could see over a radius of many miles well enough to describe the terrain and the enemy fortifications and to detect changes of position or movements of enemy forces. Under favorable conditions the balloon offered many advantages over the ordinary cavalry patrol or reconnaissance-in-force, and an examination of the reports of these observers will convince the most skeptical of their value. On the other hand, bad weather—such as high winds, rainstorms, fogs, low clouds, winter's sleet—could make the balloon utterly useless.

The most important of the Unionist aeronauts was Theodore S. C. Lowe, who organized a "balloon corps" for service with the Army of the Potomac. Though overambitious, jealous of rivals and persistently ignorant of military procedure, he was energetic, inventive, resourceful and a competent observer. Though he seems to have won the confidence of McClellan and other high officers, he was frequently in trouble with the lesser ones. He and his little organization were shoved around from one service branch to anothcr—from the topographical engineers to the quartermaster's bureau, to the engineers and finally to the signal service—and were made welcome in none. Six months after McClellan's dismissal, Lowe resigned and his "balloon corps" was disbanded. Dr. Haydon has examined a great mass of hitherto unused material_s in pre- pa ring this volume and he has made his story fresh and interesting. Adding greatly to this interest are the forty-live contemporary photographs and other illustrations at the back of the book.

Charles W. Ramsdell
The University of Texas

Lincoln Takes Command.

By John Shipley Tilley. (Chapel Hill: University of North Carolina Press. 1941. Pp. xxxvii, 334. $3.50.)[368]

In this forthright and provocative little volume Mr. Tilley, an Alabama lawyer, undertakes to show just how war came about between the United States and the Confederacy in 1861 and to fix the responsibility therefor. He gives particular attention to the situation at Forts Sumter and Pickens from December to April and to the policies and measures of Presidents Buchanan and Lincoln, on the one side, and of the Confederate officials, on the other. His opinion is that, though Buchanan and the secessionists could not agree on the rightful possession of the forts, they hoped for a peaceful solution of their difficulties and therefore refrained from pushing their respective claims too far; but that Mr. Lincoln had made up his mind at the beginning to hold the forts at any cost and, after he became President, deliberately brought on the war.

Practically none of the evidence adduced is new, but some of it is given a new significance. For instance, Mr. Tilley shows that Lincoln and Secretary Welles, with the expectation of beginning the war, deliberately violated the truce at Pensacola arranged between the Confederates and Buchanan's Secretaries of War and the Navy. He also

[368] This is the review exactly as it appeared in *The American Historical Review*, Vol. 47, No. 3 (April, 1942), 637-638.

proves (though it has been demonstrated before) that the garrison in Fort Sumter was not being starved by the South Carolinians, as intimated by Lincoln and asserted by various historians, but that Major Anderson was allowed to buy fresh meat and vegetables in Charleston until April 7, only five days before the attack. But Mr. Tilley is wrong in assuming that these purchases provided ample supplies. They were merely supplemental to the regular commissary rations of pork, flour, beans, sugar, coffee, etc., furnished from government stores, no more of which the Confederates would permit to be brought in. When these latter supplies should be used up, Anderson's men would have to evacuate or starve unless relieved. Mr. Tilley is also wrong in concluding that the missing Anderson letter of February 28 to Secretary Holt never existed, for there are too many contemporary evidences of its existence to leave room for doubt. On the other hand, there is nothing to sustain the statements of some historians that Anderson in that letter complained of a lack of food for his men.

The author castigates severely certain historians and biographers of Lincoln for their mistakes *in re* the Sumter affair, and he seems to be especially irritated by errors in the school textbooks, several of which he has pilloried in an appendix. But he has not himself escaped altogether, though most of his merely factual errors are of a minor character. One of them, several times repeated, dates on December 12 instead of December 21, 1860, Lincoln's letter containing his message to General Scott about holding the forts.

Many readers who may be impressed by the evidence Mr. Tilley has uncovered will be critical of the manner in which he has presented it. As a lawyer he has apparently been influenced by courtroom methods in the development of his case. He seems to assume something of the role of a state's attorney who is prosecuting one A. Lincoln, defendant, on the charge of inciting a war. At any rate, he follows rather closely the usual legal rules about the admission of evidence (even confining himself almost exclusively to official records); he presents only the case for the prosecution; he directs all his evidence against this one defendant (Seward is not even held responsible for the Pickens expedition); and he hammers repeatedly on the more incriminating bits of the record. The merits of the book are its incisive and challenging analysis of a much-disputed problem and its pungent style; its faults are that it is based on too narrow a concept of the problem, it is too partisan, and it contains too many questionable deductions. It will not settle the controversy over Lincoln's responsibility for the tragedy of Sumter, but it may stimulate renewed interest in the question.

Professor Avery Craven has contributed a brief foreword. There is no bibliography, but there is a very good index.

Charles W. Ramsdell
University of Texas

John Brown, Soldier of Fortune: A Critique. By Hill Publes Wilson. (Lawrence, Kansas: Hill P. Wilson, 1913. Pp. 450.)[369]

This volume was evidently written as a protest against the conclusions of Mr. Villard in his recent work, "John Brown, A Biography Fifty Years After," but it is based upon studies begun many years ago. Mr. Wilson holds that Villard's book, though scholarly, is fundamentally unsound because the author has constantly endeavored to explain Brown's career and to justify his acts in accordance with the traditional view, and that, in doing this, he has suppressed or neglected evidence which would have led to very different conclusions. Mr. Wilson's own conclusions are that Brown was a horse-thief in Kansas, and a military adventurer at Harper's Ferry, hoping by the aid of a slave insurrection to establish a military empire in the South. This view was reached as the result of investigations begun with the purpose of writing a eulogistic sketch of John Brown's career in Kansas.

The book will repay careful reading. Following the lead of Villard, the author reviews Brown's varied business career in 1852, and reveals a number of shady transactions with the idea of portraying the character of

---

[369] This is the review exactly as it appeared in *The Southwestern Historical Quarterly*, Vol. 17, No. 3 (Jan., 1914), 318-320.

the man. He also makes it clear that Brown showed no discernible interest in the slavery question prior to 1850 and then only incidentally. Having failed in business in 1854, the next year Brown followed five of his sons to Kansas as a settler, bringing along by request some arms for the free-state men furnished by the abolitionists. After examination of the evidence, the author declares that Brown took no conspicuous part as a free-state leader; but that, discouraged by the gloomy outlook for farming, he plotted to steal horses, organized a small band for that purpose, committed the murders on the Pottawatomie to cloak the theft, and exchanged the horses thus acquired for "fast running horses from Kentucky." As proof of Brown's sense of guilt in this, he always denied participation in the crime. So far from taking a prominent part in the warfare with the pro-slavery men, Brown was present at only two engagements, Black Jack and Osawatomie, in both of which he was overtaken while endeavoring to get away with stolen horses and cattle. He even left Lawrence on the eve of an expected attack by the pro-slavery forces (September 14, 1856).

Brown's campaign in the East, October, 1856, to November, 1857, for funds with which to equip a company of men for warfare in Kansas, Mr. Wilson stigmatizes as a "colossal graft upon free-state sentiment," the more palpable because conditions in Kansas were becoming peaceful. Though he raised the funds, Brown did nothing in Kansas except to make a raid into Missouri for more plunder.

About this time Brown conceived the plan that carried him to Harper's Ferry two years later. Believing that a slave insurrection would be easy to start, he began training a band of his former confederates, men of desperate character, for the conquest of the South. He plotted to seduce United States soldiers from their allegiance, and drew up a provisional constitution for his proposed conquests, which was adopted by a convention of his followers in Canada.

The fiasco at Harper's Ferry was due to the failure of the slaves to rise. Here Villard is taken severely to task for total misapprehension of Brown's plans, which Mr. Wilson thinks were not ill-advised except for the reliance upon the negroes. Brown's courage after capture, his concealment of his real plans, and his assumption of the attitude of a martyr, together with the state of the public mind resulting from the Civil War, have beclouded the memory of his crimes and selfish aims, and built up the tradition which envelopes his name. In this a series of eulogistic biographers have played their part.

Mr. Wilson has without question made out a strong case for the prosecution. At times he weakens it by making too much of uncertain evidence and by sundry harsh criticisms of Mr. Villard for the omission of material that must have seemed to the latter unimportant or irrelevant. But, on the whole, it is a very salutary corrective for much of the customary laudatory twaddle about John Brown, and it will have to be reckoned with by the students of the subject.

Charles W. Ramsdell

Part Five

Bibliography of Ramsdell's Writings

This section includes everything listed in "A Bibliography of the Writings of Charles W. Ramsdell" in the back of his book, *Behind the Lines in the Southern Confederacy* (Baton Rouge: Louisiana State University Press, 1944), as well as additional book reviews discovered by the compiler.

A BIBLIOGRAPHY OF THE WRITINGS OF CHARLES W. RAMSDELL

I. BOOKS WRITTEN

Reconstruction in Texas. 324 pp. *Studies in History, Economics and Public Law,* edited by the Faculty of Political Science of Columbia University, Vol. XXXVI, No. 1 (New York: Columbia University, Longmans, Green and Company, 1910).

A School History of Texas (with Eugene C. Barkerand Charles S. Potts). 384 pp. (Chicago: Row, Peterson and Company, 1912).

Behind the Lines in the Southern Confederacy. The Walter Lynwood Fleming Lectures in Southern History at Louisiana State University, 1937. 136 pp. (Baton Rouge: Louisiana State University Press, 1943).

II. BOOKS EDITED

The History of Bell County, by George W. Tyler. xxiii, 425 pp. (San Antonio: Naylor Company, 1936).

Laws and Joint Resolutions of the Last Session of the Confederate Congress (November 7, 1864-M arch 18, 1865) Together with the Secret Acts of Previous Congresses. xxvii, 183 pp. (Durham: Duke University Press, 1941).

III. PUBLISHED ARTICLES AND ESSAYS

"Martin McHenry Kenney," in the *Quarterly of the Texas State Historical Association,* X (1906-1907), 341-42.

"Texas from the Fall of the Confederacy to the Beginning of Reconstruction," *ibid.,* XI (1907-1908), 199-219.

"Presidential Reconstruction in Texas," *ibid.,* 277-317.

"Texas in the Confederacy, 1861-1865," in *The South in the Building of the Nation,* 13 vols. (Richmond: Southern Historical Publication Society, 1909-1913), III, 402-17.

"Texas in the New Nation, 1865-1909," *ibid.,* 417-47.

"The Last Hope of the Confederacy-John Tyler to the Governor and Authorities of Texas," in the *Quarterly of the Texas State Historical Association,* XIV (1910-1911), 129-45.

"The Frontier and Secession," in *Studies in Southern History and Politics,* inscribed to William Archibald Dunning (New York: Columbia University Press, 1914), 61-79.

"Internal Improvement Projects in Texas in the Fifties," in the *Proceedings of the Mississippi Valley Historical Association,* IX (1915-1918), 99-109.

"The Confederate Government and the Railroads," in the *American Historical Review,* XXII (1916-1917), 794-810.

"Historical Election Ballots," in the *Southwestern Historical Quarterly,* XXIII (1919-1920), 308-309.

"The Control of Manufacturing by the Confederate Government," in the *Mississippi Valley Historical Review,* VIII (1921-1922), 231-49.

"The Texas State Military Board, 1862-1865," in the *Southwestern Historical Quarterly,* XXVII (1923-1924), 253-75.

"The Preservation of Texas History," in the *North Carolina Historical Review,* VI (1929), 1-16.

"The Natural Limits of Slavery Expansion," in the *Mississippi Valley Historical Review,* XVI (1929-1930), 151-71. Reprinted in the *Southwestern Historical Quarterly,* XXXIII (1929-1930), 91-111.

"Early Chapters in the History of the (Texas) Interscholastic League," in the *Interscholastic Leaguer,* XIV (1930), No. 3.

"General Robert E. Lee's Horse Supply, 1862-1865," in the *American Historical Review,* XXXV (1929-1930), 758-77.

"The Southern Heritage," in William T. Couch (ed.), *Culture in the South* (Chapel Hill: University of North Carolina Press, 1934), 1-23.

"Some Problems Involved in Writing the History of the Confederacy," in the *Journal of Southern History,* II (1936), 133-47.

"One Hundred Years of Progress in Texas, 1836-1936," a preface to *Texas Centennial Edition, Encyclopaedia Britannica* (New York: Encyclopaedia Britannica Company, 1936).

"The Changing Interpretation of the Civil War," in the *Journal of Southern History,* III (1937), 3-27.

"Lincoln and Fort Sumter," *ibid .,* 259-88.

"Carl Sandburg's Lincoln," in the *Southern Review,* VI (1940-1941), 1941 summer, 439-53.

IV. UNPUBLISHED PAPERS

"The United States Indian Policy in Texas and the Public Lands," read before the Mississippi Valley Historical Association, Madison, Wisconsin, April 15, 1921.

"The Problem of Public Morale in the Confederacy," read before the joint session of the American Historical Association and the Mississippi Valley Historical Association, Richmond, Virginia, December, 1924.

"Barker as a Historian" (manuscript in the University of Texas Library, 1926).

"How Slavery Came to Texas," read before the East Texas Historical Association, Huntsville, March 1, 1927.

"Materials for Research in the Agricultural History of the Confederacy," read before the joint session of the American Historical Association and the Agricultural Historical Society, Chapel Hill, North Carolina, December 31, 1929.

"Was There a Reasonable Probability That the Election of Lincoln Meant an Attack upon the Institution of Slavery Within the States?" read before the Mississippi Valley Historical Association, Chattanooga, Tennessee, April 25, 1930.

V. CONTRIBUTIONS TO BIOGRAPHICAL AND HISTORICAL DICTIONARIES

A. *Dictionary of American Biography,* 20 vols. and index (New York: Charles Scribner's Sons, 1928-1937), edited by Allen Johnson and Dumas Malone.

Baker, Daniel, I, 517.
Baker, William Mumford, I, 527.
Baylor, Robert Emmet Bledsoe, II, 77-78.
Bell, Peter Hansborough, II, 160-61.
Burleson, Edward, III, 286-87.
Burleson, Rufus Clarence, III, 287-88.
Burnet, David Gouverneur, III, 292-94.
Coke, Richard, IV, 278-79.
Culberson, Charles Allen, IV, 585-86.
Culberson, David Browning, IV, 586-87.
Gorgas, Josiah, VII, 28-30.
Lubbock, Francis Richard, XI, 480-81.
Memminger, Christopher Gustavus, XII, 527-28.
Myers, Abraham Charles, XIII, 375-76.
Neighbors, Robert Simpson, XIII, 407-408.

B. *Dictionary of American History,* 5 vols. and index (New York: Charles Scribner's Sons, 1940), edited by James Truslow Adams and R. V. Coleman.

Army, Confederate, I, 109-10.
Confederate States of America, The, II, 9-13.

VI. BOOK REVIEWS

A. *American Historical Review*

John Bach McMaster, *A History of the People of the United States During Lincoln's Administration* (New York: D. Appleton and Company, 1927), XXXIII (1927-1928), 156-58.

John Gibbon, *Personal Recollections of the Civil War* (New York: G. P. Putnam's Sons, 1928), XXXIV (1928-1929), 367-68.

J. Frank Dobie, *A Vaquero of the Brush Country* (Dallas: Southwest Press, 1929), XXXV (1929-1930), 679-80.

Ella Lonn, *Salt as a Factor in the Confederacy* (New York: Walter Neale, 1933), XXXIX (1933-1934), 753-54.

James Truslow Adams, *America's Tragedy* (New York: Charles Scribner's Sons, 1934), XL (1934-1935), 529-31.

John Shipley Tilley, *Lincoln takes Command.* (Chapel Hill: University of North Carolina Press, 1941), XLVII (April, 1942), 637-38.[370]

B. *American Political Science Review*

Jesse T. Carpenter, *The South as a Conscious Minority, 1789-1861: A Study in Political Thought* (New York: New York University Press, 1930), XXV (1931), 466-68.

---

[370] Added by the compiler.

Francis Butler Simkins and Robert Hilliard Woody, *South Carolina During Reconstruction* (Chapel Hill: University of North Carolina Press, 1932), XXVI (1932), 757.

C. Historical Outlook [371]

James Ford Rhodes, *History of the United States from Hayes to McKinley, 1877-1896* (New York: Macmillan Company, 1919), XI (1920), 204.

D. *History Teacher's Magazine*

Winfred Trexler Root and Herman Vandenburg Ames, *Syllabus of American Colonial History, from the Beginning of Colonial Expansion to the Formation of the Federal Union* (New York: Longmans, Green and Company, 1912), V (1914), 64.

John Bach McMaster, *A History of the People of the United States, from the Revolution to the Civil War,* Vol. VIII, 1853-1861 (New York: D. Appleton and Company, 1913), V (1914), 97.

John W. Burgess, *The Administration of President Hayes* (New York: Charles Scribner's Sons, 1916), VIII (1917), 279-80.

[371] A continuation of the *History Teacher's Magazine.*

E. *Journal of Southern History*

Douglas Southall Freeman, *R. E. Lee: A Biography,* 4 vols. (New York: Charles Scribner's Sons, 1934-1935), I (1935), 230-36.

J. G. Randall, *The Civil War and Reconstruction* (Boston: D. C. Heath and Company, 1937), IV (1938), 532-33.

Carl Russell Fish, *The American Civil War: An Interpretation* (New York: Longmans, Green and Company, 1937), IV (1938), 533-34.

Avery Craven, *The Repressible Conflict, 1830-1861* (University, La.: Louisiana State University Press, 1939), V (1939), 553-54.

John P. Dyer, *"Fightin' Joe" Wheeler* (Baton Rouge: Louisiana State University Press, 1941), VIII (May, 1942), 279-81.[372]

F. *Mississippi Valley Historical Review*

James A. B. Sherer, *Cotton as a World Power* (Frederick A. Stokes Company, 1916), IV, (Sept., 1917), 234-36.[373]

Frank Lawrence Owsley, *State Rights in the Confederacy* (Chicago: The University of Chicago Press, 1925), XIV (June, 1927), 107-110.[374]

Haywood J. Pearce, Jr., *Benjamin H. Hill, Secession and Reconstruction* (Chicago: University of Chicago Press, 1928), XVI (1929-1930), 124-25.

---

[372] Added by the compiler.

[373] Added by the compiler.

[374] Added by the compiler.

Ulrich Bonnell Phillips, *Life and Labor in the Old South* (Boston: Little, Brown and Company, 1929), XVII (1930-1931), 160-63.

William Best Hesseltine, *Civil War Prisons: A Study in War Psychology* (Columbus: Ohio State University Press, 1930), XVII (1930-1931), 480-81.

Eric William Sheppard, *Bedford Forrest, The Confederacy's Greatest Cavalryman* (New York: Dial Press, 1930), XVIII (1931-1932), 95-96.

Robert Selph Henry, *The Story of the Confederacy* (Indianapolis: Bobbs-Merrill Company, 1931), XVIII (1931-1932), 587-88.

Annie Carpenter (Mrs. W. F.) Love, *History of Navarro County* (Dallas: Southwest Press, 1933), XX (1933-1934), 304.

Arthur Charles Cole, *The Irrepressible Conflict, 1850-1865. A History of American Life,* Vol. VII (New York: Macmillan Company, 1934), XXI (1934-1935), 279-81.

George Fort Milton, *The Eve of Conflict: Stephen A. Douglas and the Needless War* (Boston: Houghton Mifflin Company, 1934), XXII (1935-1936), 105-106.

Robert McElroy, *Jefferson Davis: The Unreal and the Real,* 2 vols. (New York: Harper and Brothers, 1937), XXV (1938-1939), 426-28.

G. *North Carolina Historical Review*[375]

Dwight Lowell Dumond, compiler, *Southern Editorials on Secession* (New York: The Century Company, 1931), X (October, 1933), 330-331.

Dwight Lowell Dumond, compiler, *The Secession Movement, 1860-1861* (New York: The Macmillan Company, 1931), X (October, 1933), 331-333.

H. *Southwestern Political and Social Science Quarterly*

Charles Howard McIlwaine, *The American Revolution: A Constitutional Interpretation* (New York: Macmillan Company, 1923), V (1924-1925), 282-84.

Charles Seymour, *The Intimate Papers of Colonel House Arranged as a Narrative,* Vols. I and II (Boston: Houghton Mifflin Company, 1926), VII (1926-1927), 307-10.

Don C. Seitz, *The Dreadful Decade, Detailing Some Phases in the History of the United States from Reconstruction to Resumption, 1869-1879* (Indianapolis: Bobbs-Merrill Company, 1926), VII (1926-1927), 326-27.

James G. Randall, *Constitutional Problems under Lincoln* (New York: D. Appleton and Company, 1926), IX (1928-1929), 357-59.

Albert J. Beveridge, *Abraham Lincoln, 1809-1858,* 2 vols. (Boston: Houghton Mifflin Company, 1928), X (1929-1930), 432-36.

---

[375] The two Ramsdell reviews in *The North Carolina Historical Review* were added by the compiler.

Charles Seymour, *The Intimate Papers of Colonel House Arranged as a Narrative.* Vols. III and IV (Boston: Houghton Mifflin Company, 1928), XI (1930-1931), 96-99.

I. *Southwestern Social Science Quarterly* [376]

Minnie Clare Boyd, *Alabama in the Fifties: A Social Study* (New York: Columbia University Press, 1931), XIII (1932-1933), 87-88.

Edith Gittings Reid, *Woodrow Wilson: The Caricature, the Myth and the Man* (New York: Oxford University Press, 1934), XV (1934-1935), 366-67.

F. Stansbury Haydon, *Aeronautics in the Union and Confederate Armies, With a Survey of Military Aeronautics Prior to 1861*, Vol. 1 (Baltimore: The Johns Hopkins Press, 1941), XXII (Sept., 1941), 187-88.[377]

J. *Southwestern Historical Quarterly*[378]

Oliver Morton Dickerson, *American Colonial Government, 1696-1765* (Cleveland: Arthur H. Clark Company, 1912), XVI (1912-1913), 214-17.

Winfred Trexler Root, *The Relations of Pennsylvania with the British Government, 1696-1765* (New York: D. Appleton and Company, 1912), XVI (1912-1913), 214-17.

---

[376] A continuation of the *Southwestern Political and Social Science Quarterly.*

[377] Added by the compiler.

[378] A continuation of the *Texas Historical Association Quarterly.*

William E. Dodd, *Statesmen of the Old South, or From Radicalism to Conservative Revolt* (New York: Macmillan Company, 1911), XVI (1912-1913), 332-33.

Edward Channing, Albert Bushnell Hart, and Frederick J. Turner, *Guide to the Study and Reading of American History* (Boston: Ginn and Company, 1912), XVI (1912-1913), 334.

Ernest William Winkler (ed.), *Journal of the Secession Convention of Texas, 1861* (Austin: Texas Library and Historical Commission, 1912), XVI (1912-1913), 430-31.

James Albert Woodburn, *The Life of Thaddeus Stevens* (Indianapolis: Bobbs-Merrill Company, 1913), XVII (1913-1914), 93-95.

Jubal Anderson Early, *Lieutenant General Jubal Anderson Early, C. S. A. Autobiographical Sketch and Narrative of the War Between the States* (Philadelphia: J. B. Lippincott Company, 1912), XVII (1913-1914), 95-96.

Hill Publes Wilson, *John Brown, Soldier of Fortune: A Critique* (Lawrence, Kan.: Hill P. Wilson, 1913), XVII (1913-1914), 318-20.

Charles King, *The True Ulysses S. Grant* (Philadelphia: J. B. Lippincott Company, 1914), XVIII (1914-1915), 420-22.

Paul Leland Haworth, *America in Ferment* (Indianapolis: Bobbs-Merrill Company, 1915), XIX (1915-1916), 207.

Carl Lotus Becker, *Beginnings of the American People. The Riverside History of the United States,* Vol. I (Boston: Houghton Miffiin Company, 1915), XIX (1915-1916), 313-14.

Frederic L. Paxson, *The New Nation. The Riverside History of the United States,* Vol. IV (Boston: Houghton Miffiin Company, 1915), XIX (1915-1916), 316.

Carter Godwin Woodson, *The Education of the Negro Prior to 1861: A History of the Education of the Colored People of the United States from the Beginning of Slavery to the Civil War* (New York: G. P. Putnam's Sons, 1915), XIX (1915-1916), 440-41.

James Sprunt, *Chronicles of Cape Fear River, 1660-1916* (Raleigh: Edwards and Broughton Printing Company, 1916), XXI (1917-1918), 424-25.

Anson Mills, *My Story* (Washington: Author, 1918), XXII (1918-1919), 200-202.

Archibald Henderson, *The Conquest of the Old Southwest* (New York: Century Company, 1920), XXV (1921-1922), 222-24.

Albert Burton Moore, *Conscription and Conflict in the Confederacy* (New York: Macmillan Company, 1924), XXIX (1925-1926), 240-43.

John Donald Wade, *Augustus Baldwin Longstreet, A Study of the Development of Culture in the South* (New York: Macmillan Company, 1924), XXIX (1925-1926), 243-44.

Louis Martin Sears, *John Slidell* (Durham: Duke University Press, 1925), XXX (1926-1927), 156-57.

E. Merton Coulter, *William G. Brownlow: Fighting Parson of the Southern Highlands* (Chapel Hill: University of North Carolina Press, 1937), XLII (1938-1939), 153-55.

Louise Biles Hill, *Joseph E. Brown and the Confederacy* (Chapel Hill: University of North Carolina Press, 1939), XLIII (1939-1940), 262-64.

Ella Lonn, *Foreigners in the Confederacy* (Chapel Hill: University of North Carolina Press, 1940), XLIV (1940-1941), 148-50.

Arndt M. Stickles, *Simon Bolivar Buckner: Borderland Knight* (Chapel Hill: University of North Carolina Press, 1940), XLIV (1940-1941), 150-51.

J. Winston Coleman, Jr., *Slavery Times in Kentucky* (Chapel Hill: University of North Carolina Press, 1940), XLIV (1940-1941), 517-19.

K. *Texas Historical Association Quarterly*

John Rose Ficklen, *History of Reconstruction in Louisiana (through 1868)* (Baltimore: Johns Hopkins Press, 1910), XIV (1910-1911), 76-78.

J. B. Polley, *Hood's Texas Brigade, Its Marches, Battles, and Achievements* (New York: Neale Publishing Company, 1910), XV (1911-1912), 90-91.

August Santleben, *A Texas Pioneer* (New York: Neale Publishing Company, 1910), XV (1911-1912), 91-92.

Finis

Made in the USA
Coppell, TX
02 November 2020